HOMOTOPY INVARIANTS IN DIFFERENTIAL GEOMETRY

by

TADASHI NAGANO

University of Notre Dame
Notre Dame, Indiana

Key Words and Phrases

Thom class, Euler class, Lefschetz number, Thom form, Poincaré dual, de Rham group, Thom isomorphism, intersection number, Blaschke formula, the localization theorem for vector fields.

International Standard Book Number 0-8218-1800-7
Library of Congress Catalog Number 52-42839
AMS 1968 Subject Classifications
5730, 5731, 5732 and 5734

Printed in the United States of America

Introduction

In this article, we will discuss the Euler class, the intersection number, the Lefschetz number, and so forth. Underlying all these notions is that of the Thom class (Thom [18]) and so we begin with its definition (Def. 1.1). Our aim of discussing these well-known concepts is to describe them in a "computable" way; indeed, we will express them with differential forms or integrals of forms, which in principle can be constructed directly from the given data, based on the *relative* de Rham theory (See Appendix for this theory). For instance, in §7 we give formulae (Theorems 7.19 and 7.23) to compute the Euler class of an oriented bundle from given data about the behavior around the vanishing point set of a given section; here we generalize various theorems including the classical Hopf formula. And in §8, among other things, the Poincaré dual of a submanifold A of a compact oriented manifold M is expressed by a closed form in terms of A itself. We note here that we do not need complete knowledge of either the cohomology group $H^*(M)$ or the dual complex of M to describe the Poincaré dual as in orthodox topology. After the definition of "the Thom form" (which represents the Thom class) we explain its basic properties along with those of the Euler class. Next our concern is following Chern's line ([4] among others), to give an explicit expression for the Thom form (Theorem 6.2), of which the Gauss-Bonnet formula should be an immediate consequence (Corollary 6.3), though the proof is carried out only when the fiber is even-dimensional. In particular, the expression will be another proof of existence of the Thom form. For that purpose, we first define connection in §2, in such a way that its existence will be evident. In §3, we explain the classifying space, viz., the Grassmann manifold mainly to introduce notation as well as to describe a certain geometric aspect of its structure, which also should be known, for a later use. In §4 we will outline the proof of Theorem 6.2. §5 will be devoted to a Thom form of

---

Partially supported by the National Science Foundation under Grant GP 7403.

Received by the editors August 18, 1969.

A part of this article was read at the 664th meeting of the American Mathematical Society on April 5, 1969 with the title: The Thom class of a vector bundle and relevant localization theorems in differential geometry.

the canonical bundle over the classifying space. The well-informed reader could thus skip §§2, 3 and 5 except for notation. However, the rest of the paper is independent of the particular expression in Theorem 6.2. In §7 we prove localization theorems for the Euler class $e(E)$. The arguments are topological except at a crucial step where we use the differentiability (or more precisely the nondegeneracy) of the given section of E to construct important vector bundle C_p. In §8, we explain the classical notions mentioned above in terms of forms; in addition, we establish localizations theorems for the Lefschetz number of a smooth map. The last §10 is a remark to our first part [13]. We have added the Appendix for an exposition of the relative de Rham theory for the reader's convenience, although this theory is not quite new either (see Leray [9] p. 112 ff). Finally, a few words of caution: we sometimes confuse forms with the cohomology classes which they represent. Our cohomology groups are all over the reals, and we work in the C^∞ category. The compactness assumption on the manifold is made for the sake of simplicity but is not an absolute necessity. Also the theory could be generalized to include manifolds with boundary as in Bott-Chern [3].

§1. Thom forms and Euler forms

Let E be a smooth oriented vector bundle over a manifold M with the projection $\pi = \pi_E \colon E \longrightarrow M$. When we have in mind a metric of E (i.e., a norm on each fiber), we use the following notations:

$$E_{[\epsilon]} = \{X \in E \mid \|X\| \leq \epsilon\}, \quad \epsilon \geq 0,$$

$$E_{(\epsilon)} = \{X \in E \mid \|X\| < \epsilon\}, \quad \epsilon > 0,$$

$$\partial E_{[\epsilon]} = \partial E_{(\epsilon)} = \text{the boundary} = E_{[\epsilon]} - E_{(\epsilon)},$$

and

$$E_x = \text{the fiber } \pi^{-1}(x) \text{ of } E \text{ over } x \in M.$$

By the fiber dimension of E we mean the common dimension of the fibers E_x of E, which will be denoted by n for most of this paper.

Definition 1.1. A form $\mathfrak{U}$ on a smooth oriented vector bundle E is called a *Thom form* of E if (1) $\mathfrak{U}$ is closed, (2) the degree of $\mathfrak{U}$ is n = the fiber dimension, (3) the support $|\mathfrak{U}|$ of $\mathfrak{U}$ is contained in some $E_{(\epsilon)}$ and (4) the integral $\int_{E_x} \mathfrak{U}$ over each fiber E_x equals one.

When $n = 0$ (so that $E = M$), let us understand that $\mathfrak{U}$ is the constant function 1 (or -1 depending on the orientation). The Thom form is a representative of the Thom class (see Thom [18] or Husemoller [5]). A few remarks are due. $\mathfrak{U}$ in turn determines the orientation. $\mathfrak{U}$ represents a cohomology class in $H^n(E, E - E_{[0]})$; neither the metric nor the particular value of ϵ is essential in the definition of $\mathfrak{U}$. When M is connected, condition (4) follows both from (4) itself with "each" replaced by "some" (and from the other conditions (1)–(3)).

Theorem 1.2. *There exists a Thom from $\mathfrak{U}$ for E and its cohomology class is unique.*

Here the uniqueness means that for any two Thom forms $\mathfrak{U}$, $\mathfrak{V}$ of E there exists an $(n-1)$-form $\mathfrak{W}$ with support $|\mathfrak{W}|$ in some $E_{(\epsilon)}$ such that $\mathfrak{U} - \mathfrak{V} = d\mathfrak{W}$. This theorem is of course well known and was proved by Thom in much greater generality; nonetheless we want to give a fairly complete proof. Until a Thom form is shown to exist (§6), we will assume its existence; its uniqueness, however, is proved in this section (see remark after the proof of Theorem 1.3).

Let $\{W_\lambda\}$ be a cell division (into a CW-complex) of M, whose cells W_λ are submanifolds (see Milnor [11]). Then there exists a cell division of $E_{[\epsilon]}$ which contains $\pi^{-1}(W_\lambda) \cap E_{(\epsilon)}$ for all λ, where $\pi: E \longrightarrow M$ is the projection. We use this in the proof of

Theorem 1.3 (The Thom isomorphism). *The map $\psi: H^k(M) \longrightarrow H^{k+n}(E_{[\epsilon]}, \partial E_{[\epsilon]})$: $\alpha \longrightarrow \pi^*\alpha \wedge \mathfrak{U}$ is an isomorphism for all $k \in \mathbf{Z}$, where the support $|\mathfrak{U}|$ of the Thom form $\mathfrak{U}$ is assumed to be contained in $E_{[\epsilon]}$. (ψ is really an isomorphism $H^*(M) \longrightarrow H^*(E, E - E_{[0]})$ and is called the Thom isomorphism.)*

Proof. α is thought of as a closed k-form on M. The map ψ is well defined since $\pi^*\alpha \wedge \mathfrak{U}$ is closed and has the support in $E_{(\epsilon)}$. The theorem is trivial unless $0 \leq k \leq \dim M$, since on the one hand all the cells $\pi^{-1}(W_\lambda) \cap E_{(\epsilon)}$ defined above are of dimension $\in [n, n + \dim N]$ and on the other hand a form $\alpha \neq 0$ has degree $\in [0, \dim M]$. Assume $0 \leq k \leq \dim M$. We proceed to show ψ is injective. Suppose there exists a cycle C of M, a linear combination of W_λ's, for which $\int_C \alpha \neq 0$. Then $\pi^{-1}(C) \cap E_{(\epsilon)}$ is a cycle of $(E_{[\epsilon]}, \partial E_{[\epsilon]})$ and the integral of $\pi^*\alpha \wedge \mathfrak{U}$ over this cycle equals $(\int_C \alpha) \cdot \int_{E_x} \mathfrak{U} = \int_C \alpha \neq 0$. Thus $\psi(\alpha) \neq 0$ in the cohomology group. ψ is moreover surjective. For let β be a closed $(n+k)$-form with support $|\beta|$ in $E_{(\epsilon)}$. Let α be the unique k-form on M such that $\int_W \alpha = \int_{\pi^{-1}(W)} \beta$ for any singular smooth k-simplex W of M. (To be more precise, we specify the orientation for $\pi^{-1}W$. For the smooth map $W: I^k \longrightarrow M$, $I = [0, 1]$, we note that the induced bundle $W^\# E$ over I^k is trivial. Fix the decomposition

$W^{\#}E \cong I^k \times \mathbf{R}^n$. Define $f_W\colon I^k \times \mathbf{R}^n \longrightarrow W^{\#}E \longrightarrow E$ in a natural way. Then $\int_{\pi^{-1}(W)}\beta$ means $\int_{I^k \times \mathbf{R}^n} f^*\beta$, which is independent of the decomposition above and is certainly convergent because $|\beta| \subset E_{[\epsilon]}$.) Then α is closed; in fact, for any singular smooth $(k+1)$-chain A, we have

$$\int_A d\alpha = \int_{\partial A}\alpha = \int_{\pi^{-1}(\partial A)}\beta = \int_{\partial(\pi^{-1}A \cap E_{[\epsilon]})}\beta$$

$$= \int_{\pi^{-1}A \cap E_{[\epsilon]}} d\beta = \int 0 = 0$$

by $|\beta| \subset E_{[\epsilon]}$ and Stokes' theorem. We claim $\psi(\alpha)$ is cohomologous to β in $H^{k+n}(E, E - E_{(\epsilon)})$. This follows since every chain of $(E, E - E_{(\epsilon)})$ with our cell division is a linear combination of $\pi^{-1}(W_\lambda)$'s and we have

$$\int_{\pi^{-1}(W_\lambda)}\psi(\alpha) = \int \pi^*\alpha \wedge \mathfrak{U} = \int_{\pi^{-1}(W_\lambda)}\alpha = \int_{\pi^{-1}(W_\lambda)}\beta.$$

Hence ψ is surjective and therefore ψ is an isomorphism. Q.E.D.

Remark. (Proof of the uniqueness). From the above argument, the uniqueness of the Thom class follows easily. Let $\mathfrak{U}$ and $\mathfrak{V}$ be two Thom forms of E, which we may assume to have the supports in $E_{[\epsilon]}$. Then the map $\beta \longmapsto \alpha$ used above $(= \psi^{-1})$ sends both $\mathfrak{U}$ and $\mathfrak{V}$ to the same 0-cocycle of M (which is the constant function one). Thus these $\mathfrak{U}$, $\mathfrak{V}$ belong to the same cohomology class. And we can now define the *Thom class* of E by the cohomology class which a Thom form represents.

The following basic properties of the Thom forms (Proposition 1.4 and 1.5) are immediate from their definition and their uniqueness.

Proposition 1.4 (Naturality). *If $f\colon E \longrightarrow F$ is a bundle map which induces on each fiber an orientation-preserving linear isomorphism (in particular if E is the pullback of F by a map between the base manifolds) then the Thom class of E is given by the pullbact $f^*\mathfrak{U}$ of a Thom form $\mathfrak{U}$ of F.*

An application: If $f\colon E \longrightarrow E$ is given by $f(X) = cX$ with a constant $c > 0$ and if $\mathfrak{U}$ is a Thom form of E with $|\mathfrak{U}| \subset E_{[\epsilon]}$ then $f^*\mathfrak{U}$ is a Thom form with $|f^*\mathfrak{U}| \subset E_{[\epsilon/c]}$. Let us note that the conclusion of Proposition 1.4 is valid under weaker conditions e.g., if f, restricted to each fiber E_x, has the mapping degree $= 1$, (i.e. $\int_{E_x} f^*\mathfrak{U} = 1$) (here f need not be linear or diffeomorphic).

Proposition 1.5. *Given two oriented vector bundles E, F over the same M, the Thom class of the Whitney sum $E \oplus F$ is given by the exterior (= wedge)* *product* $\mathfrak{U} \wedge \mathfrak{V}$ *of the Thom forms* $\mathfrak{U}$, $\mathfrak{V}$ *of* E, F *respectively, provided* $E \oplus F$ *is suitably oriented.*

Here the suitable orientation is of course defined by the following: if the bases $(a_1, \cdots, a_m)$, $(b_1, \cdots, b_n)$ for E_x, F_x, $x \in M$ are compatible with the respective orientations, then so is the $(a_1, \cdots, a_m, b_1, \cdots, b_n)$ of $(E \oplus F)_x$. We will orient $E \oplus F$ always in this fashion. ($F \oplus E$ may have the other orientation.)

Now we proceed to the Euler class, which is derived directly from the Thom class. Since the sections v: $M \to E$ of a vector bundle E are homotopic to each other, the pullbacks $v^*\mathfrak{U}$ of a Thom form define one and the same cohomology class $\in H^n(M)$. In particular, this class is obtained by inducing $\mathfrak{U}$ on $M = E_{[0]}$.

Definition 1.6. The cohomology class $v^*\mathfrak{U} \in H^n(M)$ above is called the *Euler class* of E and denoted by $e(E)$.

The following basic properties of the Euler class are known to characterize the correspondence: $E \mapsto e(E)$ up to a constant multiple (viz. along with a normalization, say, $e(T(S^2)) = 2 \cdot 1$, where $T(S^2)$ is the tangent bundle and 1 is the fundamental class $\in H^2(S^2)$; see Example below Corollary 6.4) (by Kobayashi-Nomizu [8, Chapter XII, § 5]).

Proposition 1.7 (Naturality). *If* E *is induced from* F *by a map* f *between the base manifolds, then* $e(E) = f^*e(F)$. (2) $e(E \oplus F) = e(E) \wedge e(F)$ *provided* $E \oplus F$ *is suitably oriented.*

Proof. Obvious from Propositions 1.4 and 1.5. A geometric use of $e(E)$ is illustrated by

Proposition 1.8. $e(E) = 0$ *if* E *has a nonvanishing section.*

Proof. Let v be a nonvanishing section. Put $w = 3v/\|v\|$. Let $\mathfrak{U}$ be a Thom form with support $|\mathfrak{U}|$ in $E_{[2]}$. Then $e(E) = w^*\mathfrak{U} = 0$, since $w(M) \subset E - E_{[2]}$.

Remark. The converse is known to be true if $n \geqq \dim M$ (but is in general false when $n < \dim M$). (See, e.g., Husemoller [5].)

Lemma 1.9. *Let* $\mathfrak{U}$ *be a Thom form of* E *and* ψ: $H^*(M) \to H^*(E, E - E_{[0]})$ *be the Thom isomorphism (Theorem 1.3). Then for any closed form* γ *on* E *and any section* v: $M \to E$, *one has* $\psi(v^*\gamma) = \gamma \wedge \mathfrak{U}$ *in* $H^*(E, E - E_{[0]})$.

Proof. By the definition of ψ, we have $\psi(v^*\gamma) = \pi^*v^*\gamma \wedge \mathfrak{U} = (v \circ \pi)^*\gamma \wedge \mathfrak{U}$. And $v \circ \pi$ is homotopic to the identity of E, say by the homotopy:

$X \longrightarrow tX + (1-t)v \circ \pi(X)$, $t \in [0, 1]$. Thus $\psi(v^*\gamma)$ is cohomologous with $\gamma \wedge \mathfrak{U}$.

Proposition 1.10. *The Thom isomorphism sends the Euler class $e(E)$ to $\mathfrak{U} \wedge \mathfrak{U}$. In particular $e(E) = 0$ if the fiber dimension is odd.*

Proof. This is a corollary to Lemma 1.9; put $\gamma = \mathfrak{U}$ and recall $e(E) = v^*\mathfrak{U}$. When n is odd, we have $\mathfrak{U} \wedge \mathfrak{U} = -\mathfrak{U} \wedge \mathfrak{U}$ for the n-form $\mathfrak{U}$ by the anticommutativity of the exterior product.

§2. Connections

Let E be a vector bundle over M. Then its *first jet bundle* J^1E is by definition the vector bundle over M consisting of restrictions, denoted by $j_x^1 v$, of the differentials v_* of the sections $v: M \longrightarrow E$ to the points x of M. (The components of $j_x^1 v$ are those of v together with its first derivatives at x.) The projection $j_x^1 v \longrightarrow v(x)$ gives the short exact sequence:

$$0 \longrightarrow E \otimes T(M)^* \longrightarrow J^1E \longrightarrow E \longrightarrow 0, \tag{2.1}$$

where $T(M)^*$ is the cotangent bundle of M. A *connection* of E is by definition a splitting of (2.1), or equivalently a Whitney sum decomposition $J^1E \cong E \oplus E \otimes T(M)^*$. (This definition is due to Bott in an unpublished manuscript and is equivalent to any other reasonable definition. Compare Bott-Chern [3].) Since a splitting always exists, so does a connection.

(2.2) *When a connection is given, any section $v: M \longrightarrow E$ gives rise to a vector bundle homomorphism $\nabla_V: T(M) \longrightarrow E$, where $T(M)$ is the tangent bundle of M.*

Indeed a connection is a splitting: $J^1E \longrightarrow E \otimes T(M)^*$ of (2.1) and hence converts a section of J^1E into that of $E \otimes T(M)^*$, which is naturally interpreted as a homomorphism $T(M) \longrightarrow E$. On the other hand a section v of E gives rise to a section: $x \longmapsto j_x^1 v$ of J^1E and thus ∇_v in (2.2) is defined. The homomorphism ∇_v is called the *covariant derivative* of v with respect to the connection.

A connection of E gives rise to a splitting of the short exact sequence:

$$0 \longrightarrow \pi_E^{\#}E \longrightarrow T(E) \longrightarrow \pi_E^{\#}T(M) \longrightarrow 0, \tag{2.3}$$

where $\pi^{\#}E$, etc. are the bundles induced by the projection $\pi: E \longrightarrow M$. The way our connection gives the splitting is as follows. This time we regard the connection, a splitting of (2.1), as a monomorphism $\Delta: E \longrightarrow J^1E$. For $X \in E$, the monomorphism $\Delta: E \longrightarrow J^1E$. For $X \in E$, the monomorphism: $(\pi^\# T(M))_X \longrightarrow T(E)_X$ is defined by its image which shall be the image of $\Delta(X)$: $T(M)_{\pi(X)} \longrightarrow T(E)_X$. In this paper the epimorphism: $T(E) \longrightarrow \pi^\# E$ given by this splitting of (2.3) will be tentatively called the *vertical projection* of the connection.

When E is given a metric, the (orthonormal) frame bundle P by definition consists of the orthonormal bases $p: \mathbf{R}^n \longrightarrow E_x$, $x \in M$. An orthogonal transformation $g \in O(n)$ of $\mathbf{R}^n$ acts on P by sending $p \in P$ to the composite $p \circ g = pg$, and hence on $P \times \mathbf{R}^n$ by $(p, X) \longrightarrow (pg, g^{-1}X)$. We write G for $O(n)$. G becomes a transformation group of $P \times \mathbf{R}^n$ and E is then the orbit space $(P \times \mathbf{R}^n)/G$, also denoted by $P \times {}_G R^n$. The canonical projection: $P \times \mathbf{R}^n \longrightarrow E$ is given by $(p, X) \longrightarrow p(X)$. In this way P becomes the principal G-bundle associated with E. The action $\alpha: G \times P \longrightarrow P$ extends to the action $\alpha_*: T(G) \times T(P) \longrightarrow T(P)$ of the Lie group $T(G)$, which is its differential. In particular $G \subset T(G)$ acts on P by α_* (Kobayashi [6]). The orbit space $T(P)/G$ is a vector bundle over M in a natural fashion. The projection $\pi_P: P \longrightarrow M$ induces an epimorphism: $T(P)/G \longrightarrow T(M)$ via the differential π_{P*}. If K is its kernel, then we obtain the short exact sequence:

$$(2.4) \qquad 0 \longrightarrow K \longrightarrow T(P)/G \longrightarrow T(M) \longrightarrow 0.$$

A *connection* of P is by definition a splitting of (2.4) (see Kobayashi-Nomizu [8] for the theory of connections. Also compare Kobayashi [6]). Again the existence is obvious by the same reason.

The kernel K above is the vector bundle $P \times {}_G\mathbf{G}$, where $\mathbf{G}$ is the tangent space $T(G)_e$ at the identity e, which can be identified with the space of all the skew-symmetric $n \times n$-matrices. Each $g \in G$ acts on $\mathbf{G}$ by the adjoint operation $A \longmapsto gAg^{-1}$. Thus $P \times {}_G\mathbf{G}$ is defined. Since $\mathbf{G}$ consists of linear endomorphisms of $\mathbf{R}^n$, we have a linear map: $\mathbf{G} \otimes \mathbf{R}^n \longrightarrow \mathbf{R}^n$. This map gives rise to "an action" of K on E: $K \otimes E \longrightarrow E$, which is defined for any vector bundle E associated with P.

A connection, interpreted as an epimorphism: $T(P)/G \longrightarrow K = P \times {}_G\mathbf{G}$ can be lifted up to $\omega: T(P) \longrightarrow P \times \mathbf{G} \longrightarrow \mathbf{G}$, where the second arrow indicates the canonical projection. ω is a matrix-valued one-form $(\omega_j^i)_{1 \le i,j \le n}$ on P with $\omega_i^j = -\omega_j^i$. Since ω has come from the map: $T(P)/G \longrightarrow K$, ω satisfies the following compatibility (or equivariance) condition:

$$(2.5) \qquad \omega(Yg) = g^{-1}\omega(Y)g \quad \text{for } Y \in T(P),\ g \in G.$$

And the following formula reflects the fact that the map $T(P)/G \longrightarrow P \times_G \mathbf{G}$ is splitting

$$\omega(pZ) = Z \quad \text{for } p \in P, \; Z \in \mathbf{G}. \tag{2.6}$$

Conversely it is obvious that a form ω: $T(P) \longrightarrow G$ satisfying the conditions (2.5) and (2.6) comes from some connection of P, which is unique. The form ω is called the *connection form* for the connection.

Next we shall see how a connection of P determines that of E. First $P \times \mathbf{R}^n$ is a principal G-bundle over E by the evaluation map: $P \times \mathbf{R}^n \longrightarrow E$. The analog for this principal bundle of the epimorphism: $T(P)/G \longrightarrow T(M)$ in (2.4) is:

$$T(P \times \mathbf{R}^n)/G \longrightarrow T(E). \tag{2.7}$$

On the other hand, it is a trivial matter to verify that $T(P \times \mathbf{R}^n)/G$ is naturally isomorphic with $\pi^{\#}(T(P)/G \oplus E)$ and hence we have a natural injection:

$$\pi^{\#}(T(P)/G) \longrightarrow T(P \times \mathbf{R}^n)/G. \tag{2.8}$$

With these in mind we regard a connection of P as a splitting: $T(M) \longrightarrow T(P)/G$ of (2.4) and we lift this up to a monomorphism: $\pi^{\#}T(M) \longrightarrow \pi^{\#}(T(P)/G)$, which, composed with (2.8) and (2.7), gives rise to a homomorphism: $\pi^{\#}_E T(M) \longrightarrow T(E)$. The last homomorphism assigns to $X \in E$ a linear homomorphism of $T(M)_{\pi(X)}$ into $T(E)_X$, and it is easy to see that this is a member of J^1E. Finally, the resulting map: $E \longrightarrow J^1E$ is a splitting of (2.1). Thus, a connection of E is constructed.

Now we like to express the *vertical projection* in terms of the connection form ω. So assume P has a section s: $M \longrightarrow P$, since locally this is always the case. Since then we have $T(P)/G \cong s_* T(M) \oplus K$ and $K \cong M \times \mathbf{G}$, the connection map: $T(M) \longrightarrow T(P)/G$ sends Z to $s_*(Z) - \omega(s_*Z)$. Also $E = \{s(z)\mathbf{x} \mid z \in M, \; x \in \mathbf{R}^n\}$ and hence $T(E) = \{s_*(Z)\mathbf{x} + s(z)\mathbf{y} \mid Z \in T(M)_z, \; z \in M,$ $\mathbf{x}$ and $\mathbf{y} \in \mathbf{R}^n\}$. The connection constructed above is confused with the map: $\pi^{\#}T(M) \subset E \times T(M) \longrightarrow T(E)$: $(s(z)x, Z) \longrightarrow s_*(Z)x - s^*\omega(Z)x$. Therefore the vertical map: $T(E) \longrightarrow \pi^{\#}E$ sends $s_*(z)x + s(z)y$ to $s^*\omega(Z)x + s(z)y$. We summarize this in the following:

Lemma 2.9. *Given a local frame s: $U \subset M \longrightarrow P$ defined on an open set U of M, the vertical projection is expressed with*

$$dx + (s^*\omega)x = \left[dx^i + \sum_{j=1}^{n} (s^*\omega)^i_j X^j \right]_{1 \le i \le n}, \tag{2.10}$$

on $E|U = U \times \mathbf{R}^n$.

$(s^*\omega)^i_j$ will be called the *components of the connection* with respect to s. We explain the use of (2.10) in a later section. There E will be oriented so that it makes sense to consider the bases $s \in P$ of the fibers of E which are compatible with the orientation. Then the structure group G of P is reduced from $O(n)$ to $SO(n)$. $SO(n)$ leaves invariant the n-form $X^1 \wedge \ldots \wedge X^n$ over $\mathbf{R}^n$. This implies that, θ^i denoting $dX^i + \Sigma_{j=1}^n (s^*\omega)^i_j X^j$ above, the form $\theta^1 \wedge \theta^2 \wedge \cdots \wedge \theta^n$ defines a global n-form on E which is independent of the local frame.

In order to construct another global form later on, we define the curvature form $\mathbf{K}$. First, we define a differential operator D acting on E-valued differential forms θ on M. When the degree of θ is zero so that θ is a section $u\colon M \longrightarrow E$, we define $D\theta = Du$ by $\nabla_v\colon T(M) \longrightarrow E$ in (2.2). And when θ is of the type $u \otimes \alpha$ with a section u and an ordinary form α on M, we define $D\theta = D(u \otimes \alpha)$ by $Du \wedge \alpha + u \otimes d\alpha$. D is completely and well defined by these conditions. Then it is easy to see that there exists a unique Hom(E, E)-valued 2-form $\mathbf{K}$ on M such that we have $D \circ D\theta = \mathbf{K}\theta$ for any E-valued form θ. If the connection of E is constructed from a connection form ω of P, then $\pi_P^*\mathbf{K}$ is a G-valued 2-form given by $\pi_P^* K = d\omega - [\omega, \omega] = (d\omega^i_j - \Sigma_{k=1}^n \omega^k_i \wedge \omega^k_j)_{1 \le i,j \le n}$.

We conclude this section with a remark. Given any smooth map $f\colon L \longrightarrow M$, the induced vector bundle $f^\# E$ is naturally given a metric and $f^\# P$ becomes the associated frame bundle. Let $F\colon f^\# P \longrightarrow P$ denote the canonical homomorphism. Then $F^*\omega$ is a connection form, while $F^*\pi_P^*\mathbf{K}$ is the curvature form.

§3. Grassmann manifolds

This section is devoted to some study of the structure of a Grassmann manifold, which will be needed later.

Let $G_{n,\nu}$ denote the Grassmann manifold consisting of the oriented n-dimensional vector subspaces of $\mathbf{R}^{n+\nu}$. The usual action of the special orthogonal group $SO(n+\nu)$ on $\mathbf{R}^{n+\nu}$ induces the action on $G_{n,\nu}$; thus $g \in SO(n+\nu)$ sends a subspace U of $R^{n+\nu}$ to $g(U)$ for $U \in G_{n,\nu}$. The action is transitive and it is easy to see that $G_{n,\nu}$ is in a one-to-one correspondence with $SO(n+\nu)/(SO(n) \times SO(\nu))$ together with the group action (i.e., $SO(n+\nu)$-equivariant). $G_{n,\nu}$ is given the manifold structure by which the bijection is a diffeomorphism.

(3.1) *$G_{n,\nu}$ is a homogeneous manifold $SO(n+\nu)/SO(n)\times SO(\nu)$*. We are interested in relations between $G_{n,\nu}$ and $G_{n,\nu+1}$. So we agree that $\mathbf{R}^{n+\nu}$ is contained in $\mathbf{R}^{n+\nu+1}$ as the orthogonal complement to the subspace spanned by $\mathbf{e} = \mathbf{e}_{n+\nu+1} = (0, \cdots, 0, 1)$. This inclusion induces an imbedding: $G_{n,\nu} \to G_{n,\nu+1}$ and a monomorphism: $SO(n+\nu) \to SO(n+\nu+1)$ in a natural way. Each $g \in SO(n+\nu)$ leaves $\mathbf{e}$ fixed. We denote by $N(G_{n,\nu})$ the normal bundle of $G_{n,\nu}$ imbedded in $G_{n,\nu+1}$.

The universal vector bundle (see Husemoller [5]) for the structure group $SO(n)$ is the *canonical bundle*, $E(G_{n,\nu})$, which is by definition the vector bundle over $G_{n,\nu}$ consisting of the pairs (x, U) with $x \in U \in G_{n,\nu}$. The projection sends (x, U) to U.

Lemma 3.2. *The canonical bundle $E(G_{n,\nu})$ is isomorphic with the normal bundle $N(G_{n,\nu})$ of $G_{n,\nu}$ imbedded in $G_{n,\nu+1}$.*

Proof. (We give a rather lengthy proof to explain various facts of some use. Also an elementary proof will be added.) Let V be the subspace spanned by the first n canonical basis vectors $(\mathbf{e}_i)_{1\le i\le n}$. Then $V \in G_{n,\nu} \subset G_{n,\nu+1}$. We fix V as the "origin" of these manifolds. We will introduce a local coordinate map around V for $G_{n,\nu+1}$. Let $\mathbf{G}_{n,\nu+1}$ be the vector space of all skew-symmetric linear endomorphisms α of $\mathbf{R}^{n+\nu+1}$ which send V into the orthogonal complement $V^{\perp}$ of V in $\mathbf{R}^{n+\nu+1}$. Let $p_{\nu+1}: SO(n+\nu+1) \to G_{n,\nu+1}$ be the projection which sends g to $g(V)$. Then we claim:

(3.3) *$p_{\nu+1}\circ \exp$ defines a local coordinate map from a neighborhood of 0 in $\mathbf{G}_{n,\nu+1}$ into $G_{n,\nu+1}$, where $\exp\alpha = \sum_{k=0}^{\infty}\alpha^k/k!$.*

The reason for this is as follows. Every $U \in G_{n,\nu}$ is represented uniquely by the orthogonal projection $p^U: \mathbf{R}^{n+\nu+1} \to U$. We have $p^{g(U)} = g \circ p^U \circ g^{-1}$ for $g \in SO(n+\nu+1)$. Let $g(t)$ be a smooth curve in $SO(n+\nu+1)$ with $g(0) = 1$, and f its tangent vector at $t = 0$; for instance let $g(t) = \exp(t\mathbf{g})$. Then $p^{g(t)(V)} = p_{\nu+1}(g(t))$ is a curve in $G_{n,\nu+1}$. Its tangent vector at $t = 0$ is $\mathbf{g}p^V + (p^V)^t\mathbf{g} = \mathbf{g}p^V - p^V\mathbf{g} = [\mathbf{g}, p^V]$ by $g^{-1} = {}^tg$ and $-\mathbf{g} = {}^t\mathbf{g}$. It is easy to see that $[\mathbf{g}, p^V] \neq 0$ if $\mathbf{g} \neq 0$ and $\mathbf{g} \in \mathbf{G}_{n,\nu+1}$. This means that the differential $(p_{\nu+1}\circ\exp)_*$ of $p_{\nu+1}\circ\exp$ defined on $\mathbf{G}_{n,\nu+1}$ is injective at 0. On the other hand we have $\dim \mathbf{G}_{n,\nu+1} = n(\nu+1) = \dim SO(n+\nu+1)/SO(n)\times SO(\nu+1) = \dim G_{n,\nu+1}$. Thus, (3.3) is verified. In view of (3.3), $\mathbf{G}_{n,\nu+1}$ is identified with the tangent space $T(G_{n,\nu+1})_V$ to $G_{n,\nu+1}$ at V. Naturally, we agree that

$T(G_{n,\nu})_V = \mathbf{G}_{n,\nu} = \{\alpha \in G_{n,\nu+1} | \alpha(V) \subset \mathbf{R}^{n+\nu}\}$. We introduce a norm on $G_{n,\nu+1}$ by putting $\|\alpha\|^2 = (1/2)\mathrm{Tr}(-\alpha^2)$. This norm is left invariant by $SO(n) + SO(\nu+1)$. Hence $G_{n,\nu+1}$ becomes a homogeneous Riemannian manifold with the metric defined by this norm since $T(G_{n,\nu+1})_V = G_{n,\nu+1}$ and (3.1) with $\nu + 1$ for ν. Let $\mathbf{N}$ be the orthogonal complement of $\mathbf{G}_{n,\nu}$ in $\mathbf{G}_{n,\nu+1}$. Clearly we have the legimate identifications $N(G_{n,\nu}) = \{(g(V), g(\xi)) | g \in SO(n+\nu), \xi \in \mathbf{N}\}$ and $T(G_{n,\nu}) = \{(g(V), g(\eta)) | g \in SO(n+\nu);\ \eta \in \mathbf{G}_{n,\nu}\}$ suggested by (3.1) and $T(G_{n,\nu+1})_V = \mathbf{G}_{n,\nu+1}$. Lemma 3.2 will be proved if there exists an $SO(n) \times SO(\nu)$-equivariant isomorphism of V onto $\mathbf{N}$, where one notes that the subgroup $\{1_n\} \times SO(\nu)$ acts trivially both on V and $\mathbf{N}$. We now construct such an isomorphism. Consider the map $f\colon V \to \mathrm{Hom}(\mathbf{R}^{n+\nu+1}, \mathbf{R}^{n+\nu+1})\colon x \to \xi = \mathbf{x}{}^t\mathbf{e} - \mathbf{e}{}^t\mathbf{x}$, where ${}^t\mathbf{e}$, etc. are regarded as the transposes of the matrices $\mathbf{e}$, etc. of single columns so that ξ maps $\mathbf{a}$ to $\langle \mathbf{e}, \mathbf{a}\rangle\mathbf{x} - \langle \mathbf{x}, \mathbf{a}\rangle\mathbf{e}$ expressed with the inner product $\langle\ ,\ \rangle$. Clearly ξ is skew-symmetric and orthogonal to $\mathbf{G}_{n,\nu}$ and hence ξ belongs to $\mathbf{N}$. The resulting map $f\colon V \to \mathbf{N}$ is a linear isomorphism. Finally f is $SO(n)$-equivariant, since we have $(g\mathbf{x}){}^t\mathbf{e} - \mathbf{e}{}^t(g\mathbf{x}) = g\xi g^{-1}$ for $g \in SO(n)$. Thus Lemma 3.2 is proved.

The above isomorphism of $E(G_{n,\nu})$ onto $N(G_{n,\nu})$ may be defined in a more elementary fashion, which will provide us with further information on the structure of $G_{n,\nu+1}$. Let $x(t)$, $t \in \mathbf{R}$, denote the vector space spanned by $\cos(t\|\mathbf{x}\|)\mathbf{x} - \|\mathbf{x}\|\sin(t\|\mathbf{x}\|)\mathbf{e}$ and the orthogonal complement of $\mathbf{x}$ in U for $\mathbf{x} \in U \in G_{n,\nu}$, where $\|\mathbf{x}\|$ is the length of $\mathbf{x} \in \mathbf{R}^{n+\nu}$. We have $x(0) = U$. (When $\mathbf{x} = 0$, $x(t) = U$ for all t.) The orientation of $x(t)$ is given so as to be continuous in t. Then $x(t)$ is a well-defined member of $G_{n,\nu+1}$. And $t \to x(t)$ is a smooth curve in $G_{n,\nu+1}$ which passes through U. Its initial tangent $x'(0)$ is normal to $G_{n,\nu}$ imbedded in $G_{n,\nu+1}$. The map: $(\mathbf{x}, U) \to x'(0)$ is the isomorphism in the proof above. Incidentally, the curve $x(t)$ is a geodesic in $G_{n,\nu+1}$, a fact which will not logically be used in the sequel. It is the practice to denote $x(1)$ by $\mathrm{Exp}(x'(0))$. The map $\mathrm{Exp}\colon N(G_{n,\nu}) \to G_{n,\nu+1}$ is smooth, since $x'(0) \mapsto \mathbf{x}$ is a diffeomorphism by Lemma 3.2 and $\|\mathbf{x}\|$ appears in the definition of $x(t)$ as a power series of $\|\mathbf{x}\|^2$. It is easy to see

Lemma 3.4. *$\mathrm{Exp}\colon N(G_{n,\nu}) \to G_{n,\nu+1}$ is $SO(n+\nu)$-equivariant.*

Also, it is worth noting that

Lemma 3.5. *$SO(n+\nu)$ is transitive on each sphere bundle $\partial N(G_{n,\nu})_{[\epsilon]}$.*

By digression we should like to make the following remark. The above lemmas show that the transformation group $SO(n+\nu)$ of $G_{n,\nu+1}$ has hypersurfaces as orbits. Thus [12], there exist just two singular orbits. One is $G_{n,\nu}$. The other will be interpreted as $G_{n-1,\nu+1}$ below. To put this in another way the cut locus (Kobayashi-Nomizu [8, Vol. 2]) of $G_{n,\nu}$ is $G_{n-1,\nu+1}$. In general, for a compact connected submanifold M_1 of a compact connected manifold M, M is *homeomorphic* with a compactification of the normal bundle $N(M_1)$. But $M - N(M_1)$ is not necessarily a submanifold. Fortunately in the case of $G_{n,\nu} \subset G_{n,\nu+1}$, $G_{n,\nu+1} - N(G_{n,\nu})$ is a submanifold, $G_{n-1,\nu+1}$.

To take a close look at the map Exp, we first imbed $G_{n-1,\nu+1}$ into $G_{n,\nu+1}$ by assigning to $W \in G_{n-1,\nu+1}$ the space spanned by W and $\mathbf{e}$. The image of this imbedding is exactly the points of $G_{n,\nu+1}$ which contain $\mathbf{e}$ as subspaces of $\mathrm{R}^{n+\nu+1}$. Hereafter we will denote the image by $G_{n-1,\nu+1}$.

Lemma 3.6. *Let N denote the normal bundle $N(G_{n,\nu})$ of $G_{n,\nu}$ imbedded in $G_{n,\nu+1}$. Then, through the restrictions,* Exp: $N \longrightarrow G_{n,\nu+1}$ *gives rise to a diffeomorphism: $N_{(\pi/2)} \longrightarrow G_{n,\nu+1} - G_{n-1,\nu+1}$ and a surjective map: $N_{[\pi/2]} \longrightarrow G_{n,\nu+1}$. In particular, one has the isomorphism $H^*(N, N - N_{(\pi/2)}) \cong H^*(G_{n,\nu+1}, G_{n-1,\nu+1})$.*

Proof. As for the diffeomorphism, we construct the inverse explicity. Let U_1 be an arbitrary member of $G_{n,\nu+1} - G_{n-1,\nu+1}$. By the orthogonal projection: $\mathrm{R}^{n+\nu+1} \longrightarrow \mathrm{R}^{n+\nu}$, U_1 is then mapped onto some member U of $G_{n,\nu}$. There exists a unique $\mathbf{x}$ in U such that U is $x(1)$ in the previous notation, i.e., U is spanned by $U \cap U_1$ and its orthogonal complement $(\cos \|\mathbf{x}\|)\mathbf{x} - \|\mathbf{x}\| (\sin \|\mathbf{x}\|)\mathbf{e}$.

Identifying N with $E(G_{n,\nu})$ by Lemma 3.2, we see that $U_1 \longrightarrow x'(0) = (\mathbf{x}, U)$ is the inverse of Exp; this is smooth outside $G_{n,\nu}$ and, by the implicit function theorem, in a neighborhood of $G_{n,\nu}$. Now we show the map is surjective. For any $(\mathbf{x}, U) \in E(G_{n,\nu}) = N$ with $\|\mathbf{x}\| = \pi/2$, the space $\mathrm{Exp}(\mathbf{x}, U) = x(1)$ contains $\mathbf{e}$ by the definition of $x(t)$; in other words $\mathrm{Exp}(\partial N_{[\pi/2]})$ is contained in $G_{n-1,\nu+1}$. Now $SO(n+\nu)$ is transitive on $G_{n-1,\nu+1}$; this together with Lemma 3.4, shows that $\mathrm{Exp}(\partial N_{[\pi/2]})$ coincides with $G_{n-1,\nu+1}$. Q.E.D.

§4. Outline of Proof

In view of the naturality (Proposition 1.4) and the classification theorem for the vector bundle (Husemoller [5]), to construct a Thom form of an oriented vector bundle E and in particular to prove the existence part in Theorem 1.2, we have only to work on the universal bundle $E(G_{n,\nu})$ (§3). $E(G_{n,\nu})$ is isomorphic with the normal bundle $N = N(G_{n,\nu})$ of $G_{n,\nu}$ imbedded in $G_{n,\nu+1}$

(Lemma 3.2). And $\mathrm{Exp}\colon N \to G_{n,\,\nu+1}$ maps $\partial N_{[\pi/2]}$ onto $G_{n-1,\,\nu+1}$ (Lemma 3.6). On the other hand, in case n is even, there exists a certain closed n-form Θ (see (5.7) or (5.8)) on $G_{n,\,\nu+1}$, which is to represent the Euler class of $E(G_{n,\,\nu+1})$ (Θ is the $SO(n + \nu)$-invariant form on $G_{n,\,\nu+1}$ which is unique to a constant multiple). Then Θ will turn out to vanish when Θ is induced on $G_{n-1,\,\nu+1}$. (Note that the restriction on $E|G_{n-1,\,\nu+1}$ has the nonvanishing section: $U \to (e,\, U)$.) Therefore, $\Theta \in H^n(G_{n,\,\nu+1},\, G_{n-1,\,\nu+1}) \cong H^n(N,\, N - N_{(\pi/2)})$ (see Lemma 3.6). $\mathrm{Exp}^*\Theta$, restricted to $N_{[\pi/2]}$, extends to a closed n-form $\mathfrak{U}$ on N with support in $N_{[\pi/2]}$, except that $\mathfrak{U}$ fails to be smooth on $N_{[\pi/2]}$. To recover the smoothness, we have only to use a smooth map $\phi\colon N \to N$ which is the identity on a neighborhood of $N_{[0]}$ and sends $N - N_{[\pi/2]}$ into $G_{n-1,\,\nu+1}$ so that we replace $\mathrm{Exp}^*\Theta$ with $\phi^*\mathrm{Exp}^*\Theta$. The integral of $\mathfrak{U}$ over a fiber is easily shown to be different from zero. Thus a constant multiple of $\mathfrak{U}$ will satisfy all the conditions in Definition 1.1. There still remains a technical problem. Though every bundle E over M is the pullback $f^\#E(G_{m,\,\nu})$ by some $f\colon M \to G_{n,\,\nu}$, f is not directly "computable" and hence $f^*\mathfrak{U}$ could not be called an explicit expression of a Thom form. This problem is solved simply by expressing $\mathfrak{U}$ with the curvature form and the vertical projection (§2) of the "canonical" connection of $E(G_{n,\,\nu})$ (because of the naturality of these).

In case n is odd, a slight modification for Θ is necessary. Let Θ' be the analog of Θ for $G_{n-1,\,\nu+1}$. Extend Θ' to a smooth form Θ'' on $G_{n,\,\nu+1}$. Its exterior derivative $d\Theta''$ will play the role of Θ in the even case; $d\Theta'' \in H^n(G_{n,\,\nu+1},\, G_{n-1,\,\nu+1})$ and $d\Theta''$ is the image of the standard homomorphism:

$$H^{n-1}(G_{n-1,\,\nu+1}) \to H^n(G_{n,\,\nu+1},\, G_{n-1,\,\nu+1})$$

(see Appendix). Then the pullback $\phi^*\mathrm{Exp}^*d\Theta''$ with modification by ϕ^* extends from $N_{[\pi/2]}$ to a closed n-form on N with the same support. That its integral over a fiber is not zero follows from the fact that $d\Theta''$ comes from the nonzero (in cohomology group) Θ' by the homomorphism. If the extension Θ'' is chosen so that $SO(n + \nu)$ leaves Θ'' invariant, then Θ'' hopefully can be expressed with the curvature and the vertical projection (see §2 for definition of *vertical projection*).

§5. Pullback by Exp

This section is an intermediate step in the proof of the existence theorem for Thom forms (Theorem 6.2). So our task is to find an explicit expression for the pullback $\mathrm{Exp}^*\Theta$ of a certain closed $SO(n+\nu+1)$-invariant n-form Θ defined on the Grassmann manifold $G_{n,\nu+1}$, where Exp is the exponential map $N(G_{n,\nu}) \longrightarrow G_{n,\nu+1}$ (see §4). Θ will be described explicitly later on. We must remark beforehand that n, the fiber dimension, is assumed to be even, since the Thom forms which we will construct will have quite different appearance if n is odd. For simplicity we thoroughly discuss only the even case. (When n is odd, it does not make sense to consider an $SO(n+\nu+1)$-invariant n-form, since such a form is necessarily zero.) Let $p_{\nu+1}$: $SO(n+\nu+1) \longrightarrow G_{n,\nu+1}$ be the canonical projection as before. Then the pullback $p^*_{\nu+1}\Theta$ is a left-invariant from on $SO(n+\nu+1)$. Thus, $p^*_{\nu+1}\Theta$ belongs to the exterior algebra over the real vector space of the Maurer-Cartan forms (see, e.g., Part I [13]), which is spanned by the forms $\Omega^\lambda_\mu = \Sigma^{n+\nu+1}_{\rho=1} x^\rho_\lambda dx^\rho_\mu$ at $(x^\lambda_\mu) = x \in SO(n+\nu+1)$ or the entries of the matrix $\Omega = (\Omega^\lambda_\mu) = {}^t x dx$. By the invariance of Θ along with Lemmas 3.4 and 3.5, the form Θ is uniquely determined by its value on Exp (l) where l is a one-dimensional subspace of a fiber of $N(G_{n,\nu})$. We recall that $N = N(G_{n,\nu})$ is identified with $\{(g(\xi), g(\mathbf{N}))|g \in SO(n+\nu),\ \xi \in \mathbf{N}\}$ (see §3 for $\mathbf{N}$). Let μ denote the map: $(g(\xi), g(\mathbf{N})) \longmapsto g \exp\xi$ into $SO(n+\nu+1)$ which is defined for any $\xi \in N$ and g in a small neighborhood of the identity in exp (G), as in (3.3). We then have

$$p_{\nu+1} \circ \mu = \mathrm{Exp} \tag{5.1}$$

on the definition domain of μ. Write x for $\exp\xi$. And we have $\mu^*\Omega = {}^t(g\exp\xi)d(g\exp\xi) = {}^t(gx)d(gx) = {}^t x {}^t g(dg)x + {}^t x dx$ by ${}^t gg = 1$. Putting $g = 1$, we obtain

$$\mu^*\Omega = {}^t x(dg)x + {}^t x dx \quad on \ N_V. \tag{5.2}$$

We introduce several notations for computation; $r = \|\xi\| = \|\mathbf{x}\|$, $c = \cos r$, $s = \sin r$, and assuming $\mathbf{x} \neq 0$ for the moment, $\mathbf{u} = \mathbf{x}/r$, $\mathfrak{U} = \mathbf{e}{}^t\mathbf{u} + \mathbf{u}{}^t\mathbf{e}$ and $\mathfrak{J} = \mathbf{u}{}^t\mathbf{e} - \mathbf{e}{}^t\mathbf{u}$, where we recall ξ and $\mathbf{x}$ are related by $\xi = \mathbf{x}{}^t\mathbf{e} = \mathbf{e}{}^t\mathbf{x} = r\mathfrak{J}$. A straightforward computation of $x = \exp\xi$ will give us

$$x = 1 - \mathfrak{U} + c\mathfrak{U} + s\mathfrak{J}, \tag{5.3}$$

and

$$(5.4) \qquad {}^t x dx = \mathfrak{J} dr + (c-1) d\mathfrak{U} + s d\mathfrak{J} + 2(1-c)\mathfrak{U} d\mathfrak{U},$$

or

$$(5.5) \qquad {}^t x dx = (\mathbf{e}^t\mathbf{u} - \mathbf{u}^t\mathbf{e}) dr + (1-c)(\mathbf{u}^t d\mathbf{u} - d\mathbf{u}^t\mathbf{u}) + s(\mathbf{e}^t d\mathbf{u} - (d\mathbf{u})^t\mathbf{e}).$$

To verify (5.3) and (5.4) it would be useful to note that $\mathbf{R}\mathfrak{U} + \mathbf{R}\mathfrak{J}$ is an algebra which is isomorphic with $\mathbf{C} = \mathbf{R} + \mathbf{R}i$ and that $\mathfrak{U} d\mathbf{u} = \mathfrak{J} d\mathbf{u} = 0$ by the inner product ${}^t\mathbf{uu} = 1$ and ${}^t\mathbf{ue} = 0$ and hence $\mathfrak{U} d\mathfrak{U} = \mathbf{u}^t d\mathbf{u}$, etc.

$$(5.5) \qquad \mu^*\Omega_i^\alpha = \sum_{k=1}^{n} (dg)_k^\alpha x_i^k \quad \text{on } N_V \quad \text{for } n < \alpha \leq n+\nu, \ 1 \leq i \leq n.$$

Proof. Since the α-th unit basis vector $\mathbf{e}_\alpha$, $n < \alpha \leq n + \nu$, is orthogonal to $\mathbf{e} = \mathbf{e}_{n+\nu+1}$ and $\mathbf{u} \in V$, we have $\mathfrak{U}\mathbf{e}_\alpha = 0$ and $\mathfrak{J}\mathbf{e}_\alpha = 0$, and hence $x\mathbf{e}_\alpha = \mathbf{e}_\alpha$ by (5.3). Thus $(dx)\mathbf{e}_\alpha = 0$. These two facts together with (5.2) imply

$$\mu^*\Omega_i^\alpha = {}^t\mathbf{e}_\alpha(\mu^*\Omega)\mathbf{e}_i = {}^t\mathbf{e}_\alpha({}^t x(dg)x + {}^t x dx)\mathbf{e}_i = {}^t\mathbf{e}_\alpha(dg)x\mathbf{e}^i + {}^t\mathbf{e}_\alpha(-{}^t dx)\mathbf{e}_i$$
$$= {}^t\mathbf{e}_\alpha(dg)x\mathbf{e}_i = \sum_{\lambda=1}^{n+\nu+1} (dg)_\lambda^\alpha x_i^\lambda .$$

Moreover $x\mathbf{e}_\alpha = \mathbf{e}_\alpha$ implies $x_i^\alpha = 0$ for $1 \leq i \leq n < \alpha \leq n+\nu$. We thus have $\mu^*\Omega_i^\alpha = \Sigma_{k=1}^n (dg)_k^\alpha x_i^k + (dg)_{n+\nu+1}^\alpha x_i^{n+\nu+1}$. Finally, we have $(dg)_{n+\nu+1}^\alpha = 0$ from the fact that the form dg has values in $\mathbf{G}_{n,\nu}$. (5.5) is proved. We claim

$$(5.6) \qquad \mu^*\Omega_i^{n+\nu+1} = u^i dr + s du^i \quad on \ N_V, \quad 1 \leq i \leq n,$$

where u^i is the ith component of $\mathbf{u}$.

Proof. $\mu^*\Omega_i^{n+\nu+1} = {}^t\mathbf{e}({}^t x(dg)x + {}^t x dx)\mathbf{e}_i$. Since dg has values in $\mathbf{G}_{n,\nu}$, we see that the first term ${}^t\mathbf{e}{}^t x(dg)x\mathbf{e}_i = 0$. Thus $\mu^*\Omega_i^{n+\nu+1} = {}^t\mathbf{e}{}^t x dx\mathbf{e}_i = ({}^t x dx)_i^{n+\nu+1}$, while the right-hand side equals $u^i dr + s du^i$ by (5.4′). (The right-hand side is well defined also at $r = \|\mathbf{x}\| = 0$.)

$$(5.6') \qquad \mu^*\Omega_i^{n+\nu+1} = \sum_{k=1}^{n} ((1/c)u^k dr + s du^k)x_i^k \quad on \ N_V \cap N_{(\pi/2)},$$
$$1 \leq i \leq n.$$

Proof. The formula (5.3) implies $x_i^k = \delta_i^k + (c-1)u^k u^i$, from which follows

(5.6′) by (5.6), $\Sigma u^k u^k = 1$ and $\Sigma (du)^k u^k = 0$.

Now what we have had in mind is the form Θ which represents the "Euler class" of $E(G_{n,\nu+1})$, which is given explicitly up to a universal constant multiple by

$$p^*_{\nu+1}\Theta = (\det(\Phi^i_j)_{i\leq i,j\leq n})^{1/2}, \tag{5.7}$$

i.e.

$$p^*_{\nu+1}\Theta = (1/m!2^m)\Sigma(\operatorname{sgn}\sigma)\Phi^{\sigma(1)}_{\sigma(2)}\wedge\ldots\wedge\Phi^{\sigma(k-1)}_{\sigma(k)}\wedge\ldots\wedge\Phi^{\sigma(n-1)}_{\sigma(n)}, \tag{5.8}$$

where $\Phi^i_j = \Sigma^{n+\nu+1}_{\lambda=n+1}\Omega^\lambda_i \wedge \Omega^\lambda_j m = n/2$, $\operatorname{sgn}\sigma$ is the sign of the permutation σ of $\{1, 2, \cdots, n\}$ and the summation in (5.8) ranges over all σ (see for example Kobayashi-Nomizu [8, vol. II, p. 304]).

If we put

$$k^i_j = \sum_{\alpha=n+1}^{n+\nu} (dg)^\alpha_i \wedge (dg)^\alpha_j, \quad 1 \leq i,\ j \leq n, \tag{5.9}$$

$$\alpha^i = u^i dr + s du^i, \quad 1 \leq i \leq n, \tag{5.10α}$$

and

$$\beta^i = (1/c)u^i dr + s du^i, \quad 1 \leq i \leq n, \tag{5.10β}$$

then, by (5.5), (5.6) and (5.6′), we have the matrix-valued form

$$(\Phi^i_j) = {}^t\bar{x}(k^i_j)\bar{x} + (\alpha_i \wedge \alpha_j) = {}^t\bar{x}\{(k^i_j) + (\beta^i \wedge \beta^j)\}\bar{x},$$

where $\bar{x}$ is the submatrix $(x^i_j)_{1\leq i,j\leq n}$ of x. We thus conclude

$$\mu^*\Theta = (\det({}^t\bar{x}(k^i_j)\bar{x} + (\alpha_i \wedge \alpha_j))^{1/2}_{1\leq i,j\leq n}, \tag{5.11α}$$

and

$$\mu^*\Theta = c(\det((k^i_j) + (\beta^i \wedge \beta^j))_{1\leq i,j\leq n})^{1/2}, \tag{5.11β}$$

since $\det \overline{x} = c$. Here we note that (5.11β) is valid on N_V (not only on $N_V \cap N_{(\pi/2)}$, as it should be, since each term of the right-hand side, expresses as the sum of monomials, contains dr at most once and each $1/c$ in β_i is eventually cancelled by the first factor c in the right-hand side of (5.11β). Furthermore, both (5.11α) and (5.11β) make sense and are valid on the whole $N = N(G_{n,\nu})$, where u^i is understood as the ith component of $\mathbf{u} = \mathbf{x}/\|\mathbf{x}\|$ with respect to any orthonormal basis, compatible with the orientation, of the fiber of N which contains $\mathbf{x} \in N$. This fact follows by observing the right-hand sides of (5.11α and β), consider on N_V, are invariant under the action of $SO(n)$.

Finally, we seek an expression of $\mu^*\Theta$ in terms of the quantities which are relevant to a connection of N in order to find its analog for any oriented vector bundle E. Now the (oriented orthonormal) frame bundle of $N(G_{n,\nu}) = E(G_{n,\nu})$ is the Stiefel manifold $P = SO(n+\nu)/SO(\nu)$. Each Ω^i_j, $1 \leq i, j \leq n$, is well defined on P [13]. The matrix-valued form $(\Omega^i_j)_{1 \leq i,j \leq n}$ on P is the connection form of what we call the *canonical connection* of P. The maps

$$G_{n,\nu} \overset{\mathrm{Exp}}{\longleftarrow} \mathbf{G}_{n,\nu} \overset{\exp}{\longrightarrow} SO(n+\nu) \overset{\mathrm{Proj}}{\longrightarrow} SO(n+\nu)/SO(\nu) = P$$

give rise to a local frame defined on a neighborhood of V in $G_{n,\nu}$. In terms of this frame the components of the connection are precisely $\mu^*\Omega^i_j$ restricted to $N[0] = G_{n,\nu}$. Thus we put $x = 1$ in (5.2). We then obtain $\mu^*\Omega^i_j = (dx)^i_j$, which equals zero by (5.4$'$). In other words the vertical projection at $V \in G_{n,\nu}$ (Lemma 2.9) is given by dx^i with respect to this local frame. The curvature form is

$$\mu^*(d\Omega^i_j - \sum_{k=1}^{n} \Omega^k_i \wedge \Omega^k_j),$$

which equals

$$\mu^*(\sum_{\alpha=n+1}^{n+\nu} \Omega^\alpha_i \wedge \Omega^\alpha_j) = (k^i_j)$$

on $N[0]$ by the structure equation for the Maurer-Cartan form. But (k^i_j) in (5.11α and β) is the pullback of the curvature form (k^i_j) on $N[0]$ by the projection of $N = N(G_{n,\nu})$ onto $G_{n,\nu}$. Therefore, $\mu^*\Theta$ is given by (5.11α and β) with the above interpretation of the curvature form and the understanding that

$$\alpha^i = x^i d\|\mathbf{x}\|/\|\mathbf{x}\| + \sin\|\mathbf{x}\|\,\rho^*\omega^i, \tag{5.12α}$$

and

$$(5.12\beta) \qquad\qquad \beta^i = x^i d\|\mathbf{x}\|/(\|\mathbf{x}\|\cos\|\mathbf{x}\|) + \sin\|\mathbf{x}\|\rho^*\omega^i,$$

where ω^i is the vertical projection of the canonical connection, ρ is the map: $N(G_{n,\nu}) - G_{n,\nu} \to N(G_{n,\nu}): (\mathbf{x}, U) \to (\mathbf{x}/\|\mathbf{x}\|, U)$ and where we agree that $\alpha^i = dx^i$ and $\beta^i = dx^i$ on $N(G_{n,\nu})[_0]$.

§ 6. Conclusion of Proof

In this section we finish the proof of Theorem 6.2. By first we have to introduce some notation. Let $f: \mathbf{R} \to \mathbf{R}$ be a smooth function such that (1) $f(x) = 1$ on a neighborhood of $(-\infty, 0]$ and (2) $x \to xf(x)$ is a nondecreasing function which attains its maximum value $\pi/2$ at $x = \pi$. Let E be the given oriented vector bundle with fiber dimension n. We fix a metric $\|\ \|$ on E along with a compatible metric connection of E. Define $\phi: E \to E$ by $\phi(X) = f(\|X\|)X$. Let s be an arbitrary local frame of E. Then we define one-forms α^i and β^i, $1 \le i \le n$, on the open set in E, on which s is valid, by

$$(6.1\,\alpha) \qquad\qquad \alpha^i = \rho(X)^i d\|X\| + (\sin\|X\|)\rho^*\omega^i, \text{ and}$$

$$(6.1\beta) \qquad\qquad \beta^i = \rho(X)^i d\|X\|/\cos\|X\| + (\sin\|X\|)\rho^*\omega^i$$

both at X, where ρ is the map: $E - E[_0] \longrightarrow E: X \to X/\|X\|$ with the convention that $\alpha^i = \beta^i = dX^i$ at $X = 0$ and where X^i and ω^i are the ith components of X and of the vertical projection of the connection $((\omega^i = dX^i + \Sigma\omega_j^i X^j), \S 2)$ respectively with respect to s (compare (5.12α and β)). α^i and β^i are smooth on their definition domain. Let $\mathbf{K}$ denote the curvature form which is regarded as an $n \times n$-matrix-valued 2-form by means of s. As before we denote the projection: $E \to M$ by π_E (but not by π to distinguish it from the number π). $|S^n|$ will denote the volume of the unit sphere S^n of dimension n. And finally, $\bar{x} = \bar{x}(X)$ will denote the symmetric linear endomorphism of the fiber which contains X such that (1) $\bar{x}(X) = (\cos\|X\|)X$ and (2) $\bar{x}$ is the identity on the orthogonal complement of X in the fiber. Now we are in a position to state

Theorem 6.2. *Let E be an oriented vector bundle with even fiber dimension n. Fix a metric $\|\ \|$ and a compatible metric connection for E. Then a Thom form $\mathfrak{U}$ of E with support in $E[_\pi]$ is given by $\mathfrak{U} = \phi^*\mathfrak{U}'$ with $\mathfrak{U}'$ defined by*

$$(6.3\alpha) \qquad\qquad \mathfrak{U}' = (2/|S^n|)\,(\det({}^t\bar{x}(\pi_E^*\mathbf{K})\bar{x} + (\alpha_i \wedge \alpha_j)_{1 \le i,j \in n})^{1/2},$$

or equivalently by

(6.3β) $\mathfrak{U}' = (2/|S^n|)\cos\|X\|(\det(\pi_E^* \mathbf{K} + (\beta_i \wedge \beta_j)_{1 \le i,j \le n}))^{1/2}$.

It should be obvious that $\mathfrak{U}'$ and hence $\mathfrak{U}$ are independent of the local frame s and defined globally on E, since the determinant is invariant under the orthogonal transformations. (See (5.7) and (5.8) for the symbol det.)

Corollary 6.4 (The Gauss-Bonnet formula). *The Euler class of E is*:

$$e(E) = (2/|S^n|)(\det \mathbf{K})^{1/2}.$$

(Kobayashi-Nomizu [8,vol.2,p.318]; $(-1)^p$ should be deleted from the formula for $\pi^*(\nu)$ on p. 318.)

Proof. Obvious from (6.3β) and Definition 1.9.

Example. When $E = T(S^n)$ and S^n is the unit sphere, $(\det K)^{1/2}$ is the volume element of S^n. Thus $e(E) = 2.1 \in H^n(S^n)$, or $\chi(s^n) = 2$.

Remark 6.5. Theorem 6.2 implies the existence (Theorem 1.2) of the Thom class in the case where n is even.

Remark 6.6. As remarked in § 1, it is a trivial matter to obtain a Thom form with support in $E_{[\epsilon]}$ for any $\epsilon > 0$ from $\mathfrak{U}$ above.

Now we begin the proof. According to Definition 1.1, we must show the following:

(6.7) $\mathfrak{U}$ *is of degree* n,

(6.8) *The support* $|\mathfrak{U}|$ *of* $\mathfrak{U}$ *is contained in* $E_{[\pi]}$,

(6.9) *The integral of* $\mathfrak{U}$ *over each fiber equals one, and*

(6.10) $\mathfrak{U}$ *is closed.*

First (6.7) is obvious if one recalls (5.7) and (5.8). Next we prove (6.8). In view of the factor $\cos\|x\|$ in (6.3β) the form $\mathfrak{U}'$ vanishes if $\mathfrak{U}'$ is induced on $\partial E_{[\pi/2]}$. On the other hand we have $\phi(E - E_{(\pi)}) = \partial E_{[\pi/2]}$ by the definition of ϕ. Thus, $|\mathfrak{U}| = |\phi^* \mathfrak{U}'|$ is contained in $E_{[\pi]}$. If $\mathfrak{U}'$ is induced on a fiber E_x for the proof of (6.9), $\mathfrak{U}'$ becomes $(2/|S^n|)\alpha^1 \wedge \alpha^2 \wedge \ldots \wedge \alpha^n$ by (6.3α) and $\alpha^i = \rho(X)^i d\|X\| + (\sin\|X\|)\rho^* dX^i$ on E_x. There $\mathfrak{U}'$ equals $(2/|S^n|)(\sin\|X\|)^{n-1} d\|X\| \wedge \sum_{i=1}^n (-1)^{i-1} u^i du^1 \wedge \ldots \wedge du^i \wedge \ldots \wedge du^n = (2/|S^n|)(\sin\|X\|)^{n-1}$ (the volume element of E_x) $= (2/|S^n|F^n$ (the volume element of S^n), where $\mathbf{u} = \rho(X) = X/\|X\|$ and F is the map: $E_x \to S^n: X \mapsto (\sin\|X\|/\|X\|)X + \cos\|X\|\mathbf{e}$ with the identification $\mathbf{R}^{n+1} = E_x \oplus \mathbf{Re} \supset S^n$ and with $\|\mathbf{e}\| = 1$. Since $F(E_x \cap E_{[\pi/2]})$ is the upper hemisphere of S^n, the integral of $\mathfrak{U}'$ over $E_x \cap E_{[\pi/2]}$ thus equals one. (An alternative proof will go like this: in § 5, $\text{Exp}(N(G_{n,\nu})_V \cap N(G_{n,\nu})_{[\pi/2]})$ is isomorphic with the unit hemisphere and Θ, restricted to this hemisphere, can be interpreted as the volume element. The analogy between (5.12a) and (6.1a) establishes the above result.) Now the effect of ϕ on the integral is favorable, since ϕ, restricted to the regular points, is an orientation-preserving diffeomorphism as is easily seen from its definition. It remains to show (6.10). Unfortunately the author cannot give a direct proof. That is one of the reasons why he has employed somewhat cumbersome arguments in §5. We have to appeal to the universal bundle $E(G_{n,\nu})$. E is the pullback of $E(G_{n,\nu}) = N(G_{n,\nu})$ by some map of M into $G_{n,\nu}$ for a sufficiently large ν, by the classification theorem for the bundles (Husemoller [5]). The metric of E also is pullbacked at the same time. Since the connection is given beforehand too and since we have not proved that the cohomology class of $\mathfrak{U}$ is independent of the connection as in Bott-Chern [3], we need a stronger theorem of Narashman-Narasimhan [14] to the effect that the map: $M \to G_{n,\nu}$ can be chosen so that it pulls the canonical connection (§ 5) of $N(G_{n,\nu})$ back to the given connection of E. Thus we may assume that $\mathfrak{U}'$ is the pullback of $(2/|S^n|)\mu^*\Theta$ in § 5. Since Θ is closed (see, e.g., [13]), so are $\mathfrak{U}'$ and $\mathfrak{U} = \phi^*\mathfrak{U}'$. Theorem 6.2 is proved.

The odd case can be discussed in a similar way with the modification mentioned at the end of § 4.

§ 7. Localization theorems for the Euler class

Given a section v: $M \to E$, we want to describe the Euler class $e(E)$ by means of the behavior of v around (or at) the vanishing point set $F = v^{-1}(0) \subset M$ and of F under certain assumptions (Theorems 7.19 and 7.23). To be more precise, let E be a smooth oriented vector bundle over M as before and first assume:

(7.1) (Regularity). *Each component F_p of F is a smooth submanifold.*

For convenience we fix a metric connection for E. Then the covariant derivative ∇_v of v is a homomorphism: $T(M) \to E$. The dimension of the kernel $(\nabla_v|T(M)_x)^{-1}(0)$ of ∇_v at x depends on $x \in M$ in general. If it is constant, the collection $\mathrm{Ker}\,\nabla_v$ of the kernels for all x becomes a vector bundle. At any rate we have $T(F_p) \subset \mathrm{Ker}\,\nabla_v$, since $v|F_p = 0$ and hence $\nabla_v = 0$ on $T(F_p)$. We assume

$$(7.2) \qquad T(F_p) = \mathrm{Ker}\,\nabla_v \quad \text{over} \quad F_p,$$

where the right-side means $\bigcup_{x\in F_p} \mathrm{Ker}\,(\nabla_v|T(M)_x$.

Note that the condition on (7.2) is independent of the connection.

Example 7.3. Let E be the cotangent bundles $T(M)^*$. Let $f\colon M \to \mathbf{R}$ be a smooth function. Then the conditions (7.1) and (7.2) are satisfied by $v = df$ if and only if the critical points form a submanifold F and each component F_p of F is a nondegenerate critical manifold.

Example 7.4. Let $E = T(M)$. Then v be an infinitesimal automorphism of a connection of E. Then both (7.1) and (7.2) are satisfied, since the components of v are linear functions with respect to the normal coordinates $(= \operatorname{Exp}|T(M)_x)$ centered at each $x \in F$.

To obtain a clear-cut result we further assume:

(7.5) *The normal bundle $N(F_p)$ of each component F_p imbedded in M is orientable (and is given a fixed orientation).*

By (7.2) we have the short exact sequence:

$$0 \to T(F_p) \to T(M) \to \operatorname{Im} \nabla_v \to 0 \text{ over } F_p. \tag{7.6}$$

In other words, we have $\operatorname{Im} \nabla_v \cong N_p$, where $N_p = N(F_p)$. Then we have

$$0 \to N(F_p) \to E|F_p \to C_p \to 0, \tag{7.7}$$

where $C_p = \bigcup_{x \in F_p} \operatorname{Coker}(\nabla_v|T(M)_x)$. Since E is orientable, we see that the condition (7.5) is equivalent to

(7.8) *Each C_p is orientable.*

Example 7.9. When F is discrete, (7.5) is satisfied if and only if M is orientable.

Example 7.10. If v is a holomorphic section of a holomorphic vector bundle, then (7.5) is automatically satisfied under the assumptions (7.1) and (7.2).

Example 7.11. When $E = T(M)$ and v is a Killing vector field (an infinitesimal isometry) for a Riemannian metric on M, all the conditions (7.1), (7.2), and (7.6) are satisfied (see Example 7.4), since the fibers of $N(F_p)$ is then naturally given a symplectic (or complex) structure by $\nabla_v|T(M)_x$ which is skew-symmetric (Kobayashi [7]).

We finally assume for simplicity that

(7.12) *M is compact (without boundary) and connected.*

Then the F_p are compact and finite in number. Hence there is a tubular neighborhood M_p of F_p such that the closure $\overline{M}_p$ of M_p is a compact manifold with the boundary and $\overline{M}_p$ is disjoint from $\overline{M}_q$ for $q \neq p$. Therefore for some $\epsilon > 0$ each component of $v^{-1}(E_{[\epsilon]})$ is contained in some M_p. Let $\mathfrak{U}$ be a Thom form of E with $|\mathfrak{U}| \subset E[\epsilon]$. $\mathfrak{U}$ is induced on $E|M_p$ to yield a Thom form $\mathfrak{U}_p$; $|\mathfrak{U}_p| \subset (E|M_p)[\epsilon]$. We have $v^*\mathfrak{U} = \Sigma v_p^*\mathfrak{U}_p^*$, i.e.

$$(7.13) \qquad e(E) = \Sigma_p v_p^*\mathfrak{U}_p$$

where $v_p = v|M_p$ and $v_p^*\mathfrak{U}_p$ is confused with its extension to a closed form on M with the same support, viz., with its image under the natural homomorphism: $H^n(\bar{M}_p, \partial M_p) \longrightarrow H^n(M, M - M_p) \longrightarrow H^n(M)$. Now F_p is a deformation retract of M_p, or more precisely M_p has a bundle structure over F_p which is isomorphic with the normal bundle $N_p = N(F_p)$. We may assume Exp: $N(F_p)[\epsilon] \longrightarrow M_p$ is a diffeomorphism. Let π_p: $M_p \longrightarrow F_p$ be the projection. Then (7.7) gives the isomorphism:

$$(7.14) \qquad E|M_p \cong \pi_p^\# C_p \oplus \pi_p^\# N_p, \quad N_p = N(F_p).$$

By Proposition 1.5, we have the following relation between the Thom forms:

$$(7.15) \qquad \mathfrak{U}_p = \mathfrak{U}(\pi_p^\# C_p) \wedge \mathfrak{U}(\pi_p^\# N_p),$$

where C_p is oriented by this.

We decompose $v_p = v|M_p$ into $v_p = v_C + v_n$ in accordance with (7.14). Then we have

$$(7.16) \qquad v_p^*\mathfrak{U}_p = v_C^*\mathfrak{U}(\pi_p^\# C_p) \wedge v_N^*\mathfrak{U}(\pi_p^\# N_p).$$

The supports of these forms are contained in $v^{-1}(E[\epsilon]) \cap M_p$. Hence (7.16) is valid in $H^n(\bar{M}_p, \partial M_p)$. By Proposition 1.7, we have

$$(7.17) \qquad v_C^*\mathfrak{U}(\pi_p^\# C_p) = e(\pi_p^\# C_p) = \pi_p^* e(C_p).$$

$v_N^*\mathfrak{U}(\pi_p^\# N_p)$ in (7.16), is a closed form with compact support in M_p with degree equal to codim F_p. Therefore, by the uniqueness of the Thom class (Theorem 1.2), $v_N^*\mathfrak{U}(\pi_p^\# N_p)$ must be a scalar multiple $a_p\mathfrak{U}(M_p)$ of a Thom form $\mathfrak{U}(M_p)$ of M_p with the oriented vector bundle structure $\cong N_p$. Thus, ψ_p denoting the Thom isomorphism (Theorem 1.3) of the vector bundle M_p, we see from (7.16) and (7.17) that $v_p^*\mathfrak{U}_p = a_p\psi_p(e(C_p))$. Hence, (7.13) gives

$$(7.18) \qquad e(E) = \Sigma_p a_p \psi_p(e(C_p)).$$

But we claim $a_p = 1$. For the proof, let $\mathfrak{v}$ denote the composite of v_N: $M_p \longrightarrow \pi_p^\# N_p \subset E|M_p$ with the natural homomorphism: $\pi_p^\# N_p \longrightarrow N_p$. If M_p is identified with N_p, $\mathfrak{d}$ is thought of as a smooth map: $N_p \to N_p$. $\mathfrak{d}$ is the identity on $F_p = (N_p)_{[0]}$ and fiber preserving, as can be easily verified. Restricted to a neighborhood of zero in each fiber, $\mathfrak{d}$ is a diffeomorphism by the nondegeneracy condition (7.2). On the other hand, we have

$$a_p \mathfrak{U}(N_p) = v_N^* \mathfrak{U}(\pi_p^\# N_p) = \mathfrak{d}^* \mathfrak{U}(N_p). \tag{7.18a}$$

Therefore, a_p is either $+1$ or -1. It remains to show that $a_p \neq -1$, i.e., $\mathfrak{d}$ preserves the orientation. But this is trivial since the first factor v_N of v preserves it because of the way $\pi_p^\# N_p$ in $E|M_p$ or $\operatorname{Im} \nabla_v$ over F_p is oriented.

Theorem 7.19. *Let v be a smooth section of an oriented vector bundle E over a compact manifold, subject to the conditions of regularity* (7.1), *nondegeneracy* (7.2) *and orientability* (7.6). *Then the Euler class of E is given by*

$$e(E) = \Sigma_p \psi_p(e(C_p)),$$

in $H^n(M)$, where ψ_p is the Thom isomorphism (*Theorem* 1.3) *of the normal bundle* (*identified with a tubular neighborhood*) *of the connected component F_p of the vanishing point set $F = v^{-1}(0)$, C_p is the vector bundle* Coker (∇_v) *over F_p, oriented by* (7.15), *and the summation ranges over all connected components F_p.*

We like to apply this theorem to the case of $E = T(M)$. Then C_p is naturally isomorphic with $T(F_p) = \operatorname{Ker} \nabla_v$ over F_p. However, its orientation given by (7.15) may differ from that of $T(F_p)$ given by $\mathfrak{U}(T(M)_x) = \mathfrak{U}(T(F_p)_x) \wedge \mathfrak{U}(N(F_p)_x)$, $x \in F_p$. Let $i(p)$ denote $+1$ or -1 according as these orientations are the same or not. The number $i(p)$ is called the *index* of v at F_p. Note that $i(p)$ remains unchanged when the orientation of $N(F_p)$ is reversed.

Definition 7.20. For a compact, oriented manifold M the *Euler number*, $\chi(M)$, is the number $\int_M e(T(M))$. The number $\chi(M)$ is independent of the orientation of M.

Corollary 7.21. *Under the assumptions of Theorem* 7.19, *further assume that E is the tangent bundle $T(M)$. Then $\chi(M) = \Sigma_p i(p)\chi(F_p)$, where F_p is a connected component of the vanishing point set F and $i(p)$ is the index of the vector field v at F_p.* (*This corollary is briefly mentioned in Atiyah-Singer* [1] *at* p. 576.)

Proof. By Theorem 7.19 we have

$$\chi(M) = \int_M e(T(M)) = \Sigma_p \int_M \psi_p(e(C_p)) = \Sigma \int_M i(p)\psi_p(e(T(F_p))).$$

And

$$\int_M \psi_p(e(T(F_p))) = \int_M \pi_p^* e(T(M)) \wedge U(N(F_p))$$
$$= \int_{N(F_p)} = \int_{F_p} e(T(M)) = \chi(F_p). \text{ Q.E.D.}$$

Remark. A simple algorithm for $i(p)$. Let $(x^i)_{1 \leq i \leq n}$ be a local coordinate system of M around $0 \in F_p$. Assume F_p is given locally by $x^{n(p)+1} = \dots = x^n = 0$, $n(p) = \dim F_p$. Then $i(x)$ is the sign of $\det(\partial_k v^a(0))_{n(p)+1 \leq a, b \leq n}$.

Example. In the special case where $v = df$ for some function f, Corollary 7.21 implies the Morse *equality* under the orientability condition on F_p, which follows of course from the discreteness of F (provided M is orientable). (Here we have identified $T(M)^*$ with $T(M)$ by a Riemannian metric.) Indeed if $k(p)$ denotes the number of the negative eigenvalues of the Hessian $(\partial^2 f/\partial x^i \partial x^j)$ of f at any point of F_p, then clearly we have $i(p) = (-1)^{k(p)}$ so that $\chi(M) = \Sigma(-1)^{k(p)}\chi(F_p)$.

Applied to Example 7.11, Corollary 7.21 leads to a theorem of Kobayashi [7] to the effect that we have $\chi(M) = \Sigma_p \chi(F_p)$ for a Killing vector field v on a compact oriented Riemannian manifold.

Next we are interested in the case where $C_p = 0$, i.e., $N(F_p) \cong E|F_p$ by ∇_v. Again we define $i(p)$ according as to whether ∇_v preserves the orientation or not. Then $e(C_p) = i(p) \in H^0(F_p)$. Thus $\psi_p(e(C_p)) = i(p)\mathfrak{U}(N(F_p))$. We call $i(x)$ "the index" of v at F_p, which depends on the orientations of $N(F_p)$ and E. Later we will see that $\mathfrak{U}(N(F_p))$ is the Poincaré dual of F_p when M is orientable (Proposition 8.9).

Corollary 7.22. *Beside the assumptions of Theorem 7.19, assume that the fiber dimensions of all C_p are zero. Then*

$$e(E) = \Sigma_p i(p)\mathfrak{U}(N(F_p)), .$$

where $i(p)$ is the "index" of v at F_p. In particular if the fiber dimension n of E equals $\dim M$ (hence F is discrete) then $\int_M e(E) = \Sigma_p i(p)$, i.e., $i(E)$ gives the number of the vanishing points with the index counted. (*Compare Bott-Chern* [3].)

When $E = T(M)$, the nondegeneracy condition (7.2) is not necessary. Really (7.2) was used to obtain the decomposition (7.14). To find a substitute for (7.1) in the case $E = T(M)$, we have only to put $C_p = T(F_p)$, oriented by (7.15). And the arguments for the proof of Theorem 7.19 work with this modification. However the number a_p, which is now defined by (7.18α), in general is no longer 1. We denote a_p by $i(p)$ and call it the *index* of v at F_p.

With this in mind, we generalize the famous Hopf formula concerning the indices of a vector field:

Theorem 7.23. *Let v be a vector field on a compact, oriented manifold M. Assume that v satisfies the regularity condition* (7.1) *and the orientability condition* (7.5). *Then* $\chi(M) = \Sigma_p i(p)\chi(F_p)$.

Remark. Let us repeat that nondegeneracy is not assumed in this theorem.

Remark 7.24. With Theorem 7.21 we can compute $e(E)$ from the information on F and on the first jet $j^1 v$ (§ 2) at a point x_p of each F_p, whereas with Theorem 7.23 we need more information on v, viz., its behavior on a neighborhood of x_p in M (or, the germ of v at x_p).

§ 8. Some basic concepts from topology

The purpose of this section is to define such classical notions as the intersection (of cycles), the Poincaré dual and the linking coefficient in the framework of relative de Rham theory and in a "computable" way to explain some of their well-known fundamental properties (see for instance, Spanier [17] or Samelson [16] for an orthodox theory). Some results of this section will be used in the next section concerning the Lefschetz number.

Let M be an oriented manifold of dimension n. For the sake of simplicity we assume that M is compact and has no boundary. Given a compact oriented submanifold A of M, we orient the normal bundle $N(A)$ in such a way that we have the relation $\mathfrak{U}(T(M)|A) = \mathfrak{U}(T(A)) \wedge \mathfrak{U}(N(A))$ among the Thom classes (Proposition 1.5). Next we regard $\mathfrak{U}(N(A))$ as a closed form on M in the following way (and as we did in § 7). Fix a Riemannian metric on M. Then the expontential map $\mathrm{Exp}\colon N(A) \longrightarrow M$, is, restricted to some $N(A)_{(\epsilon)}$, a diffeomorphism onto a tubular neighborhood M_A of A in M. If we choose as a representative of $\mathfrak{U}(N(A))$ a Thom form with support in $N(A)_{(\epsilon)}$, then $\mathfrak{U}(N(A))$ is pulled back to the diffeomorphism by Exp^{-1}. Then we extend the form $(\mathrm{Exp}^{-1})^*\mathfrak{U}(N(A))$ on M_A to the closed form on the whole of M and having the same support. This is possible since the support $|(\mathrm{Exp}^{-1})^*\mathfrak{U}(N(A))| \subset M_A$ is compact. We denote the resulting form on M by $\mathbf{U}(A)$. $\mathbf{U}(A)$ will also denote the cohomology class $\in H^{\operatorname{codim} A}(M)$ which it represents. $\mathbf{U}(A)$ will turn out to be determined by the homology class of A.

We apply this general convention to the diagonal manifold $\Delta = \{(x, x)|x \in M\} \subset M \times M$, defining a closed n-form $\mathbf{U}(\Delta)$ on $M \times M$; the orientation of Δ is given in such a way that the first projection $\pi_1\colon M \times M \longrightarrow M\colon (x, y) \mapsto x$ preserves the

the orientation when π_1 is restricted to Δ. In other words the orientation of Δ is given by

(8.1) $$\int_{\{x\}\times M} \mathrm{U}(\Delta) = 1 \quad \text{for } x \in M.$$

Let A and B be (oriented) finite smooth singular cycles of M. Put $\dim A = p$, $\dim B = q$. We assume $p + q = n$ throughout this section unless otherwise stated.

Definition 8.2. The *intersection number*, $I(A, B)$, of A, B above in M is the real number $(-1)^p \int_{A\times B} \mathrm{U}(\Delta)$, where $p = \dim A = \operatorname{codim} B$. Note that this definition applies to the special case where compact oriented manifolds A and B with $p + q = n$ are given with smooth maps $f\colon A \longrightarrow M$, and $g\colon B \longrightarrow M$; $I(f, g) = (-1)^p \int_{A\times B} (f \times g)^* \mathrm{U}(\Delta)$. A geometric meaning of $I(A, B)$ is illustrated by Proposition 8.14, but first we like to explain some immediate properties of the intersection number.

Proposition 8.3. (1) $I(\{x\}, M) = 1 = I(M, \{y\})$ *for* $x, y \in M$,

(2) $I(B, A) = (-1)^{pq} I(A, B)$, $p = \dim A$, $q = \dim B$.

(3) I *induces a linear map*: $H_p(M) \times H_q(M) \longrightarrow \mathbf{R}$, *and*

(4) $I(A, B) = 0$ *if* A *and* B *are disjoint.*

Proof. First to show (2), we define the map $t\colon M \times M \longrightarrow M \times M\colon (x, y) \longrightarrow (y, x)$. Then we have $t^* \mathrm{U}(\Delta) = (-1)^n \mathrm{U}(\Delta)$ in $H^n(M \times M)$. Thus

$$I(B, A) = (-1)^q \int_{B\times A} \mathrm{U}(\Delta) = (-1)^{q+pq} \int_{t(A\times B)} \mathrm{U}(\Delta)$$

$$= (-1)^{q+pq} \int_{A\times B} t^* \mathrm{U}(\Delta) = (-1)^{q+pq+n} \int_{A\times B} \mathrm{U}(\Delta) = (-1)^{pq} (-1)^p \int_{A\times B} \mathrm{U}(\Delta)$$

$$= (-1)^{pq} I(A, B).$$

That $I(A, B)$ is bilinear is obvious. Hence for the proof of (3), we have only to show that if A is a bounding cycle $\partial A'$ then $I(A, B) = 0$ (use (2) when B is bounding). Then $A \times B = \partial(A' \times B)$ and thus we have

$$I(A, B) = (-1)^p \int_{\partial(A'\times B)} \mathrm{U}(\Delta)$$

$$= (-1)^p \int_{A'\times B} d\mathrm{U}(\Delta) = (-1)^p \int 0 = 0$$

by Stokes' theorem. Now the first equality of (1) follows from (8.1) and the second form (2) and the first. Finally, the assumption of (4) means that $A \times B$ is a finite cycle in $M \times M - \Delta$. Then, taking $|\mathrm{U}(\Delta)|$ sufficiently small, we can regard $A \times B$

as a cycle in $M \times M - |U(\Delta)|$, from which we immediately conclude $\int_{A\times B} \mathrm{U}(\Delta) = 0$.

Proposition 8.4. *$H_p(M)$ and $H(M)$, $p + q = n$, are dually paired by the intersection number $I(A, B)$.* (See (3) of Proposition 8.3.)

For the proof we need some preparation. Given a closed q-form α, we denote by $\phi(\alpha)$ a p-dimensional cycle (or the homology class) such that we have

(8.5) $\int_{\phi(\alpha)} \beta = \int_M \beta \wedge \alpha$ *for any closed p-form β on M.* The existence of a $\phi(\alpha)$ is an immediate consequence of the de Rham theory; really ϕ is a linear map: $H^q(M) \longrightarrow H_p(M)$.

Definition 8.6. α and $\phi(\alpha)$ are called the *Poincaré dual* of each other.

Lemma 8.7. $\int_{\phi(\alpha)\times B} \mathrm{U}(\Delta) = (-1)^p \int_B \alpha$, *and hence* $I(\phi(\alpha), B) = \int_B \alpha$.

Proof. (8.5) implies that the left-hand side equals $(-1)^{pq} \int_{M\times B} \pi_1^* \alpha \wedge \mathrm{U}(\Delta)$. Let Let π_2 denote the second projection: $M \times M \longrightarrow M$: $(x, y) \mapsto y$. Then $\pi_1^*\alpha$ and $\pi_2^*\alpha$ represent the same cohomology class of Δ when they are induced on Δ. Hence we have $\int_{M\times B} \pi_1^*\alpha \wedge \mathrm{U}(\Delta) = \int_{M\times B} \pi_2^* \alpha \wedge \mathrm{U}(\Delta)$ by Proposition 1.9. Moreover, $\pi_2^*\alpha \wedge \mathrm{U}(\Delta) = (-1)^{nq} \mathrm{U}(\Delta) \wedge \pi_2^*\alpha$. And finally we have $\int_{M\times B} \mathrm{U}(\Delta) \wedge \pi_2^*\alpha = (-1)^n \int_B \alpha$ by $\int_{M\times\{b\}} \mathrm{U}(\Delta) = (-1)^n$ for any $b \in M$, which follows from (1) of Proposition 8.3. These together give the first formula of Lemma 8.7. The second follows from the definition of I.

Proposition 8.8 (*The Poincaré duality theorem*). *The linear map ϕ: $H^{n-p}(M) \longrightarrow H_p(M)$ is an isomorphism.*

Sin ϕ is defined for all p, we think of it as a linear map $H^q(M) \oplus H^p(M) \longrightarrow H_p(M) \oplus H_q(M)$, $p + q = n$. Then the definition domain and the range of ϕ are of the same finite dimension. Thus we have only to show that this ϕ or ϕ: $H^q(M) \longrightarrow H_p(M)$ is injective. If $\phi(\alpha) = 0$ then we have $\int_B \alpha = 0$ for any q-cycle B by Lemma 8.7. Hence we have $\alpha = 0$ by the de Rham theorem, and Proposition 8.8 is proved.

Proof of Proposition 8.4. Obvious from Lemma 8.7 and Proposition 8.8; in fact, $I(A, B) = \int_B \phi^{-1}(A)$ and the right-hand side is the canonical pairing $H^q(M) \otimes H_q(M) \longrightarrow \mathbf{R}$.

The next proposition makes "computable" the Poincaré duals of submanifolds.

Proposition 8.9. *The Poincaré dual $\phi^{-1}(A) = \mathrm{U}(A)$ if A is a compact oriented submanifold.*

Proof. For any closed p-form β we have $\int_A \beta = \int_M \beta \wedge \mathrm{U}(A)$ by Proposition 1.9. Compared with (8.5), this proves the proposition.

Proposition 8.10. *$I(A, B) = \int_B \mathrm{U}(A)$ if A is a compact, oriented submanifold.*

Proof. Obvious from Lemma 8.7 and Proposition 8.9. Let M_1, M_2 be the compact, oriented submanifolds of M. We say that M_1 meets M_2 *transversally* when we have $T(M)_x = T(M_1)_x + T(M_2)_x$ for any $x \in M_1 \cap M_2$. Then $M_1 \cap M_2$ is a compact, orientable submanifold of dimension $= \dim M_1 + \dim M_2 - \dim M$. We orient $M_1 \cap M_2$ as follows. Along $M_1 \cap M_2$, we have the relation $N(M_1 \cap M_2) = N(M_1) \oplus N(M_2)$ between the normal bundles. The orientation of $M_1 \cap M_2$ is given by $\mathfrak{U}(N(M_1 \cap M_2) = \mathfrak{U}(N(M_1)) \wedge \mathfrak{U}(N(M_2))$ over each connected component of $M_1 \cap M_2$ (see Proposition 1.5). Then the next proposition will be evident from what we have done:

Proposition 8.11. *Let M_1, M_2 be compact, oriented submanifolds of M. Assume M_1 meets M_2 transversally. Let the submanifold $M_1 \cap M_2$ be oriented as above. Then one has*

$$\mathbf{U}(M_1 \cap M_2) = \mathbf{U}(M_1) \wedge \mathbf{U}(M_2), \tag{8.12}$$

or equivalently the relation

$$\phi^{-1}(M_1 \cap M_2) = \phi^{-1}(M_1) \wedge \phi^{-1}(M_2) \tag{8.13}$$

among the Poincaré duals.

We apply this to the special case where $M_1 = A$, $M_2 = B$ with $\dim A + \dim B = n$. Then $A \cap B$ is a finite set. To each point $x \in A \cap B$ we assign the number $i(x) = \pm 1$ defined by $\mathfrak{U}(T(M)_x) = i(x)\mathfrak{U}(T(A)_x) \wedge \mathfrak{U}(T(B)_x)$. The number $i(x)$ is called the *intersection index* at x. We have the following localization theorem for the intersection number by Proposition 8.11:

Proposition 8.14. *The intersection number $I(A, B) = \Sigma_{x \in A \cap B}\, i(x)$ where $i(x)$ is the intersection index ± 1 at x, if A, B are compact, oriented submanifolds of M with $\dim A + \dim B = \dim M$ and A meets B transversally.*

Example 8.15. Let M be the torus $\mathbf{R}^n/\mathbf{Z}^n$. We confuse the canonical coordinates of $\mathbf{R}^n$ with local coordinates of M. Let A and B be compact, oriented submanifolds defined by the equations $x^{p+1} = \cdots = x^n = 0$ and $x^1 = \cdots = x^p = 0$, respectively. Then we have $\mathbf{U}(A) = dx^{p+1} \wedge \ldots \wedge dx^n$ and $\mathbf{U}(B) = (-1)^{pq} dx^1 \ldots dx^p$. If $(x^1, \cdots, x^n, y^1, \cdots, y^n)$ is the canonical coordinate system for $\mathbf{R}^n \times \mathbf{R}^n$, then $\mathbf{U}(\Delta) = d(y^1 - x^1) \wedge \ldots \wedge d(y^n - x^n)$ for $\Delta \subset M \times M$. It follows $I(A, B) = 1$.

Example. In the notation of §§ 3 and 5, the Euler class $e(E(G_{n,\nu+1})) = (2/|S^n|\Theta)$ is the Poincaré dual of $G_{n,\nu}$ contained in $G_{n,\nu+1}$.

Remark 8.16. Since a tubular neighborhood of $\Delta \subset M \times M$, interpreted as the normal bundle $N(\Delta)$, is isomorphic with the oriented tangent bundle $T(M)$, we have the intersection number $I(\Delta, \Delta) = \chi(M) = (-1)^n \chi(M)$, where $\chi(M)$ is the Euler number. Of course, $\chi(M) = 0$ for an odd n (Lemma 1.10).

Remark 8.17. The intersection number $I(A, B)$ is equal to $\int_M \phi^{-1}(A) \wedge \phi^{-1}(B)$. So, more generally, $\phi(\phi^{-1}(A) \wedge \phi^{-1}(B))$ is called the *intersection (class)* of homology class A, B even if $\dim A + \dim B \neq \dim M$. Also we note that the wedge product $\alpha \Delta \beta$ of closed forms is called the *cup product* of the cohomology classes which α, β represent.

Closely related to the notion of the intersection is that of the linking coefficient (or linking number). We confine our discussion on this topic to the definition and an example. Fix a Riemannian metric on M, which we still assume to be a compact, oriented manifold. The metric is naturally extended to $M \times M$. Let $U_\epsilon(\Delta)$ be a closed form having the support contained in an ϵ-neighborhood of Δ and belonging to the Poincaré dual of Δ (so that $U_\epsilon(\Delta)$ can be regarded as a Thom form of the normal bundle $N(\Delta)$).

Definition 8.18. For disjoint cycles $A = \partial A'$ and $C = \partial C'$ with $\dim A + \dim C = \dim M - 1$, *their linking number*, $\mathrm{Lk}(A, C)$ is $\lim_{\epsilon \to 0} \int_{A' \times C} (-1)^{\dim A + 1} U_\epsilon(\Delta)$.

The limit exists since A and C are disjoint or, more precisely, since the integral is independent of ϵ when ϵ is small enough so that (the image of) $A \times C$ is disjoint from the ϵ-neighborhood of Δ, $L_k(A, C)$ is independent of the choice of A or $U_\epsilon(\Delta)$. Also $L_k(A, C)$ will not change if A varies smoothly without meeting C. In short, $L_k(A, C)$ can be interpreted as the intersection number of the *chain* A' and the cycle C. When A or C is a compact, oriented submanifold, $L_k(A, C)$ can be expressed as the integral, say $\int_{C'} U(A)$ of a Thom form of the normal bundle of the one, A.

Example (The Hopf invariant). (See Part I, [13] for the reference.) Assume $\dim M = 2q - 1$ and $H_{q-1}(M) = 0$. Let N be a compact, oriented and connected manifold of dimension q. Let 1 be the fundamental class $\in H^q(N)$. Then, for a smooth map $f \colon M \to N$, the pullback $f^* 1$ is exact by the assumption $H_q(M) = 0$; we have $f^* 1 = d\beta$ for some $(q-1)$-form β. The number $\int_M f^* 1 \wedge \beta$ is called the Hopf invariant, denoted by $H(f)$ (J.H.C. Whitehead, H. Whitney). $H(f)$ is determined by the homotopy class of f. Suppose there exists a point $y \in N$ such that $A = f^{-1}(y)$ is a $(q-1)$-dimensional submanifold and the restriction of f to a tubular neighborhood of A is a submersion (i.e., rank $f = \dim N$ everywhere). (Any smooth map: $M \to N$ is arbitrarily approximated by such a map, according to another theorem of Thom.) Take a connected neighborhood V of y in N such that $f^{-1}(V)$ is trivially fibered over V by f. Let ω be a closed form in 1 with $|\omega| \subset f^{-1}(V)$. Then $f^*\omega$ is a Thom form of A or rather a form in the Poincaré dual $U(A)$ of A. This is true for $C = f^{-1}(z)$ if z is sufficiently close to y; i.e., $f^*\omega = U(C)$. Since $H_{q-1}(M) = 0$, C is a bounding cycle: $C = \partial C'$. Then

$$H(f) = \int_M \beta \wedge f^*\omega = \int_M \beta \wedge U(C) = \int_C \beta = \int_{\partial C'} \beta = \int_{C'} d\beta = \int_{C'} U(A) = L_k(A, C),$$

which recovers the original definition by Hopf.

§ 9. The Lefschetz number and localization theorems

Let M be a compact, oriented manifold of dimension n, as in the previous section. Given a smooth map $f: M \to M$ we denote by $\mathrm{gr}(f)$ the graph manifold $\{(x, f(x)) | x \in M\} \subset M \times M$. The orientation of $\mathrm{gr}(f)$ is given so that $(x, f(x)) \to x$ preserves the orientation. Let Δ be the diagonal manifold. Δ with the orientation is the graph of the identity map; $\Delta = \mathrm{gr}(id)$.

Definition 9.1. *The Lefschetz number, $L(f)$, of a smooth map $f: M \to M$ is the intersection number $I(\mathrm{gr}(f), \mathrm{gr}(id))$ of the graphs.*

Proposition 9.2. (*The Lefschetz fixed point theorem*).

(1) $L(f) = 0$ *if f has no fixed point,*

(2) $L(id) = (-1)^n \chi(M)$, *and*

(3) $L(f)$ *is a homotopy invariant*; $L(g) = L(f)$ *if g is homotopic to f.*

Proof. (1) follows from (4) of Proposition 8.3. (2) is a consequence of Remark 8.16. And (3) results from (3) of Proposition 8.3.

We need very little modification to derive the analogs of the localization theorems in § 7 for the Lefschetz number. In fact, a tubular neighborhood M_Δ of Δ is interpreted as the tangent bundle $T(M)$ and its intersection with $\mathrm{gr}(f)$, $M_\Delta \cap \mathrm{gr}(f)$, as a section over a neighborhood of the fixed point set $F = \{x \in M | f(x) = x\}$. To be more precise, we fix a Riemannian metric on M. Restricting the exponential map $\mathrm{Exp}: T(M) \to M$ to $T(M)_{(\epsilon)}$ for a sufficiently small positive ϵ, we obtain a diffeomorphism of $T(M)_{(\epsilon)}$ onto a tubular neighborhood M_Δ of Δ. This diffeomorphism sends $X \in T(M)_{(\epsilon)} \cap T(M)_x$ to $(x, \mathrm{Exp}(X))$. Thus $M_\Delta \cap \mathrm{gr}(f)$ is interpreted as the section $\sigma_f: x \to X$ given by $\mathrm{Exp}\, X = f(x)$ and defined on a neighborhood $N(F)$ of F. On the other hand, $L(f) = I(\mathrm{gr}(f), \Delta) = (-1)^n I(\Delta, \mathrm{gr}(f)) = (-1)^n \int_{\mathrm{gr}(f)} U(\Delta)$ by (2) of Proposition 8.3 and Proposition 8.10. Hence we have

(9.3) $$L(f) = (-1)^n \int_{N(F)} \sigma_f^* \mathfrak{U},$$

where $\mathfrak{U}$ is a Thom form of $T(M)$ with $|\mathfrak{U}| \subset T(M)_{(\epsilon)}$. Therefore we can use the arguments for the proof of Corollary 7.21 or Theorem 7.19 under analogous assumptions to obtain a similar formula $L(f) = (-1)^n \Sigma\, i(p)\chi(F_p)$. The number $i(p)$ is called the *fixed point index* of f at F_p. A simple method to compute $i(p)$ is given as follows. Take a point $0 \in F_p$ and a local coordinate system $(x^i)_{1 \leq i \leq n}$. Express f in terms of this system: $f = (f^i(x))_{1 \leq i \leq n}$. Then $(f^i(x) - x^i)_{1 \leq i \leq n}$ is an approximation of the section σ_f, which is good enough for our purpose. Suppose F_p is given by $x^{n(p)+1} = \cdots = x^n = 0$, $n(p) = \dim F_p$, near 0. Then $i(p)$ is the sign ± 1 of $\det(\partial_{x^b} f^a(0) - \delta^a_b)_{n(p)+1 \leq a, b \leq n}$.

Theorem 9.4. *Let f: $M \longrightarrow M$ be a smooth map of a compact, oriented manifold M of dimension n into itself. Assume that*

(1) (*regularity and orientability*) *the fixed point set F is an orientable submanifold and*

(2) (*nondegeneracy*) *the rank of f equals* $\operatorname{codim} F_p$ *at any point of each connected component F_p of F. Then the Lefschetz number of f satisfies*

(9.5) $$L(f) = (-1)^n \Sigma\, i(p)\chi(F_p),$$

where $i(p)$ is the fixed point index of f at F_p.

Remark. When f: $M \longrightarrow M$ is an isometry for a Riemannian metric on M and $M \times M$ is given the product metric, the diagonal manifold Δ is totally geodesic and $\mathrm{gr}(f)$ is the image of Δ under the isometry $id \times f$ of the ambient manifold $M \times M$. These facts seem to mean that the local data like the curvature and the second fundamental form are of no use for computing $L(f)$.

To conclude this section, we like to show that our $L(f)$ and $\chi(M)$ are the same as the classical ones (9.8) and (9.9). Let $(\alpha^p_1, \cdots, \alpha^p_{b(p)})$ be the basis for $H^p(M)$. By the Poincaré duality theorem (Proposition 8.8), we may normalize this basis in such a way that

$$\int_M \alpha^p_i \wedge \alpha^{n-p}_j = \delta_{ij},\ 1 \leq i, j \leq b(p),\ \ p < n-p,$$

and, if n is even,

$$\int_M \alpha^{n/2}_i \wedge \alpha^{n/2}_{*j} = \epsilon(i)\delta_{ij},\ 1 \leq i, j \leq b(n/2),$$

where $*$ is an involutive permutation of $\{1, 2, \cdots, b(n/2)\}$ and $\epsilon(i) = \pm 1$. From

(8.5) and the Künneth theorem (Spanier [17]) to the effect that the map $\alpha \otimes \beta \longrightarrow \pi_1^*\alpha \wedge \pi_2^*\beta$ is an isomorphism $H^N(M) \otimes H^*(M) \longrightarrow H^*(M \times M)$, we see that the Poincaré dual $\mathrm{U}(\Delta)$ is $M \times M$ is given by

$$(9.7) \qquad \mathrm{U}(\Delta) = \sum_{0 \leq p < n/2,\, 1 \leq i \leq b(p)} (-1)^p \pi_1^* \alpha_i^p \wedge \pi_2^* \alpha_i^{n-p}$$

$$+ \sum_{0 \leq i \leq b(n/2)} (-1)^{n/2} \epsilon(i) \pi_1^* \alpha_i^{n/2} \wedge \pi_2^* \alpha_{*i}^{n/2}$$

$$+ \sum_{n/2 < p \leq n} (-1)^{p+p(n-p)} \pi_1^* \alpha_i^p \wedge \pi_2^* \alpha_i^{n-2}.$$

(We would say that the right hand side is not "computable" in our vague sense since we need the complete knowledge of $H^*(M)$ for obtaining it.) Now each $f^*\alpha_i^p$ is cohomologous to some $\sum_{j=1}^{b(p)} F_i^{p,j} \alpha_j^p$, $F_i^{p,j} \in \mathbf{R}$. Let F^p denote the matrix $(F_i^{p,j})_{1 \leq i,j \leq b(p)}$. Then we have

$$(9.8) \qquad L(f) = I(\mathrm{gr}\, f, \Delta) = \int_\Delta \mathrm{U}(\Delta) = \sum_{p=0}^{n} (-1)^p \mathrm{Tr} F^p.$$

By putting $f =$ identity, we have

$$(9.9) \qquad \chi(M) = \sum_{p=0}^{n} (-1)^p b(p),$$

since $\mathrm{Tr} F^p$ is then $b(p)$.

§ 10. The Blaschke formula

This section is a supplementary remark to Proposition 4.1 of Part I [13], which expresses the Euler class Θ of the normal bundle of a manifold immersed in an euclidean space with the exterior polynomial in the components of the second fundamental form.

Let M be a closed submanifold of a manifold L. A Thom form of the normal bundle of M extends to a closed form $\mathrm{U}(M)$ on L with the same support, as in § 8. Put $n = \operatorname{codim} M$. Assume $H^n(L) = 0$. Then $\mathrm{U}(M)$ is exact; $\mathrm{U}(M) = d\theta$ for some form θ on L. Hence the Euler class $= \iota^*\mathrm{U}(M) = \iota^* d\theta = d\iota^*\theta$ is cohomologous to zero, where ι is the inclusion map: $M \longrightarrow L$. Thus we have by the Gauss-Bonnet formula (Corollary 6.4):

Proposition 8.1. *Let M be a closed submanifold with even codimension n of* *a manifold L whose normal bundle $N(M)$ is orientable. Assume $H^n(L) = 0$. Then the cohomology class of $(\det \mathbf{K})^{1/2}$ is zero, where $\mathbf{K}$ is the curvature form of an arbitrary metric connection of $N(M)$.*

Now if L is a euclidean space, M is a submanifold with even codimension n, and $N(M)$ is given the induced metric connection, then $(\det \mathbf{K})^{1/2}$ is the form Θ in the proposition of Part I mentioned above. Thus, by Proposition 8.1, Θ is cohomologous to zero, which fact is the classical Blaschke formula $\int_M \Theta = 0$ in the case where $\dim L = 4$, $n = 2$ and M is compact, oriented.

Appendix

We will formulate and prove the relative de Rham theorem (Theorem A3) concerning the cohomology group $H^*(M, N)$ over $\mathbf{R}$ of a compact bounded manifold M relative to a compact bounded submanifold N, since the theorem is seemingly less known, though it is not quite new (Leray [9]; F. J. Flaherty's paper (to appear)). A few words should be mentioned about the proof. We will not use the usual (absolute) de Rham theorem, but rather prove this at the same time. The category on which we work will be of class C^∞. We use only the existence of a Morse function and some elementary theorems in homological algebra. (See for instance, Spanier [17] for the theorems in the sequel.) The Morse function will play the role of a substitute for the smooth triangulation. (This idea was used by Samelson [15].) We will not use the sheaf theory, since we want to obtain the idomorphism by means of the integrals of forms; this explicit isomorphism is important in differential geometry. We do not aim at the most general formulation of the theorem but at a fast development of the theory with some reasonable generality.

§ A1. The relative de Rham groups

The relative de Rham theorem will be proved in the category of bounded manifolds. But we find it convenient to define and study the de Rham groups in the category of the cornered manifolds. These generalized manifolds are defined as follows. We recall that a (usual) smooth manifold structure Σ of M is a set of homeomorphisms $\lambda\colon U_\lambda \to V_\lambda$ between open sets U_λ and V_λ in $\mathbf{R}^n$ and M respectively, subject to the condition among others to the effect that $\mu^{-1} \circ \lambda$ is smooth for $\lambda, \mu \in \Sigma$. Now a *bounded manifold* and a *cornered manifold* respectively are defined by the above with $\mathbf{R}^n$ replaced by $\mathbf{R}^{n-1} \times [0,\infty)$ and $[0, \infty)^n$, where by a smooth map we mean a map defined on an open set U in $\mathbf{R}^{n-1} \times [0, \infty)$ or $[0, \infty)^n$ into $\mathbf{R}^m$ which entends to a smooth map in the usual sense on a neighborhood of U in $\mathbf{R}^n$. The *smooth maps* between bounded or cornered manifolds are defined in the obvious way. The *boundary* of a bounded or cornered manifold M, denoted by ∂M, is by definition the subset of M consisting of all points which are in the image of the boundary $\partial(\mathbf{R}^{n-1} \times [0, \infty))$ or $\partial([0, \infty)^n)$ in $\mathbf{R}^n$ under some (hence any) coordinate map λ. Every bounded or cornered manifold can "extend beyond its boundary" to a smooth manifold, which is not at all unique. Thus, a smooth map between bounded or cornered manifolds extends to a smooth map between usual smooth manifolds. A bounded manifold is a cornered one and its boundary is a smooth manifold. A smooth manifold M is a cornered manifold with $\partial M = \emptyset$ (= the empty set). (These are characterizing properties.)

In the rest of this section, our category will consist of the following: the objects are the pairs (M, N) of compact cornered manifolds M and compact cornered submanifolds N of M and the morphisms are smooth maps $f: (M, N) \longrightarrow (M', N')$, i.e., smooth maps $f: M \longrightarrow N$ with $f(N) \subset N'$. N may have the same dimension as M. We identify $(M, \emptyset)$ with M as usual.

Next we like to define the space, $\Omega(M)$, comprising all smooth differential forms on M. For that purpose we have only to define the tangent space $T(M)_x$ at a point $x \in \partial M$. Let $T(M)_x$ be the tangent space $T(M')_x$ of a smooth extension M' of M beyond ∂M. Though M' is not unique, $T(M)_x = T(M')_x$ is uniquely determined. $T(M)$ then becomes a cornered manifold and a smooth vector bundle over M. Let $\Omega(M, N)$ denote the real vector space of all differential forms $\omega \in \Omega(M)$ on M with $\iota^*\omega = 0$, where ι is the inclusion map $N \longrightarrow M$. (ω can extend to a smooth form beyond ∂M.) $\Omega(M, N)$ is graded by the degrees of the forms; $\Omega(M, N) = \bigoplus_{p=0}^{n} \Omega^p(M, N)$ where $\Omega^p(M, N)$ consists of p-forms. The exterior derivation, d, maps $\Omega(M, N)$ into itself: $d\iota^*\omega = \iota^* d\omega = 0$. Thus $\Omega(M, N)$ is a cochain complex over the real. The cohomology group $H^p(\Omega(M, N))$ will be called the *pth de Rham group of M relative to N* and denoted by $\mathfrak{D}^p(M, N)$. A morphism $f: (M, N) \longrightarrow (M', N')$ induces the cochain map $f^*: \Omega(M', N') \longrightarrow \Omega(M, N)$; in fact $\iota^* f^* \omega' = f^* \iota'^* \omega' = 0$ and $df^* = f^* d$. Thus, f also induces a homomorphism: $\mathfrak{D}^p(M', N') \longrightarrow \mathfrak{D}^p(M, N)$, which we denote by the same f^*. We will prove that the relative de Rham groups satisfy the following axioms of a cohomology theory, as formulated by Eilenberg-Steenrod:

(1) *The assignments*: $(M, N) \mapsto \mathfrak{D}^p(M, N)$ *and* $(f: (M, N) \longrightarrow (M', N')) \mapsto (f^*: \mathfrak{D}^p(M', N') \longrightarrow \mathfrak{D}^p(M, N))$ *define a contravariant functor of our category into that of the real vector spaces for an integer* p;

(2) *To* (M, N) *there corresponds the exact sequence*

$$\cdots \longrightarrow \mathfrak{D}^{p-1}(N) \xrightarrow{\delta} \mathfrak{D}^p(M, N) \xrightarrow{\kappa^*} \mathfrak{D}^p(M) \xrightarrow{\iota^*} \mathfrak{D}^p(N) \longrightarrow \cdots$$

such that, for any $f\colon (M, N) \to (M', N')$, *the diagram below is commutative*:

$$\begin{array}{ccccccccc}
\cdots \to \mathfrak{D}^{p-1}(N) & \xrightarrow{\delta} & \mathfrak{D}^p(M, N) & \xrightarrow{\kappa^*} & \mathfrak{D}^p(M) & \xrightarrow{\iota^*} & \mathfrak{D}^p(N) & \to \cdots \\
\downarrow f^* & & \downarrow f^* & & \downarrow f^* & & \downarrow f^* & \\
\cdots \to \mathfrak{D}^{p-1}(N') & \xrightarrow{\delta} & \mathfrak{D}^p(M', N') & \xrightarrow{\kappa^*} & \mathfrak{D}^p(M') & \xrightarrow{\iota^*} & \mathfrak{D}^p(N') & \to \cdots
\end{array}$$

where ι *is as above and* κ *is the inclusion map*: $(M, \emptyset) \to (M, N)$.

(3) *If two maps* $f_0, f_1\colon (M, N) \to (M', N')$ *are homotopic, then* f_0 *and* f_1 *induce the same homomorphism* $f_0^* = f_1^*\colon \mathfrak{D}^p(M', N') \to \mathfrak{D}^p(M, N)$;

(4) $\iota^*\colon \mathfrak{D}^p(M, N) \to \mathfrak{D}^p(M - U, N - U)$ *is an isomorphism for the inclusion map* ι *of the object* $(M - U, N - U)$ *into* (M, N) *if* U *is open in* M *and its closure* $\bar{U}$ *is contained in the interior of* $N \subset M$; *and*

(5) $\mathfrak{D}^p(\{x\}) = 0$ *for* $p \neq 0$ *and* $\mathfrak{D}^0(\{x\}) = \mathbf{R}$, *where* $\{x\}$ *is a single point.*

(1) and (5) are obvious. By a well-known theorem in homological algebra, (2) follows from the fact that we have the short exact sequence:

$$(6) \qquad 0 \to \Omega(M, N) \xrightarrow{\kappa^*} \Omega(M) \xrightarrow{\iota^*} \Omega(N) \to 0.$$

We will prove this fact. Since $\Omega(M, N)$ are defined as the kernel of ι^*, we have only to show that ι^* is surjective. We first observe the case where $\partial N = \emptyset$ and $N \cap \partial M = \emptyset$. Take a tubular neighborhood M_N of N in M, which we identify with the normal bundle of N. Given a form θ on N, its pullback $\pi^*\theta$ by the projection $M_N \to N$ is a form on M_N. As usual we choose a smooth function $F\colon M_N \to [0, 1]$ such that (i) the restriction $F|N = 1$ and (ii) the support of F is a compact neighborhood of N. Then the support $|F\pi^*\theta|$ is contained in that of F. Thus $F\pi^*\theta$ extends to a smooth form μ on M with the same support. Clearly we have $\iota^*\mu = \theta$, and the surjectivity of ι^* is proved in this case. In the general case we extend M and N beyond their boundaries to smooth manifolds M' and N' with $N' \subset M'$. Then a tubular neighborhood M_N' of N' in M' and its projection onto N' are defined. θ extends first to a form on N' (if N' is sufficiently small) and then to a form on M_N'. The product of the resulting form and a function F extends to a form ω on M' (hence on M) if the support of F is a compact neighborhood $\subset M_{N'}'$ in N'. We have $\iota^*\omega = \theta$ if F is chosen beforehand so that the restriction $F|N = 1$. Therefore the sequence (6) is exact. δ is defined on the level of forms as follows; given a closed form α on N, extend α to a form β on M, then $d\beta$ gives $\delta(\alpha)$.

Now we proceed to show the homotopy invariant expressed by (3). In general $M_1 \times M_2$ is a cornered manifold provided this is true of M_1 and M_2, simply because $[0, \infty)^m \times [0, \infty)^n = [0, \infty)^{m+n}$. Given a cornered manifold M and a real number $t \in I = [0, 1]$ we denote by j_t the inclusion map $M \longrightarrow M \times \{t\} \longrightarrow M \times I$. The assumption of (3) means that there exists a smooth map $F\colon (M \times I, N \times I) \longrightarrow (M', N')$ such that $f_1 = F \circ j_1$ and $f_0 = F \circ j_0$. In order to construct a homotopy on the cochain level, we assign to each p-form ϕ on $M \times I$ a $(p-1)$-form $D\phi$ on M well defined by $\int_B D\phi = (-1)^{p-1}\int_{B\times I}\phi$ for every (finite, oriented, smooth) singular $(p-1)$-chain B of M; in terms of local coordinates (x^i) of M,

$$\phi = \Sigma f_{i_1\cdots i_p}(x, t)dx^{i_1} \wedge \ldots \wedge dx^{i_p} + \Sigma g_{j_1\cdots j_{p-1}}(x, t)dx^{j_1} \wedge \ldots \wedge dx^{j_{p-1}} \wedge dt$$

is thus sent to

$$D\phi = \Sigma\left(\textstyle\int_0^1 g_{j_1\cdots j_{p-1}}(x, t)dx\right)dx^{j_1} \wedge \ldots \wedge dx^{j_{p-1}}.$$

We will prove the following homotopy formula:

$$j_1^*\phi - j_0^*\phi = Dd\phi + dD\phi. \tag{7}$$

In fact for any singular p-chain A, we have

$$\begin{aligned}\textstyle\int_A(j_1^*\phi - j_0^*\phi) &= \textstyle\int_{j_1(A)}\phi - \int_{j_0(A)}\phi = (-1)^p\int_{A\times\partial I}\phi \\ &= Dd\phi + \textstyle\int_{\partial A} D\phi = Dd\phi + dD\phi\end{aligned}$$

by Stokes' theorem and $\partial(A \times I) = \partial A \times I + (-1)^p A \times \partial I$. Now we will complete the proof of (3). Let ω be a closed form in $\Omega(M', N')$. $F^*\omega$ is then a closed form on $M \times I$. By (7) we have $f_1^*\omega - f_2^*\omega = j_1^*F^*\omega - j_2^*F^*\omega = dDF^*\omega$. Moreover, $\iota^*DF^*\omega = D\iota^*F^*\omega = D(F \circ \iota)^*\omega = 0$. Thus we have $f_1^*\omega = f_2^*\omega$ in $\mathfrak{D}^*(M, N)$, and (3) is proved.

Finally, we must verify the excision axiom (4). But this is obvious since clearly we have the isomorphism $\Omega(M, N) \cong \Omega(M - U, N - U)$ by the inclusion map $\iota\colon (M, - U, N - U) \longrightarrow (M, N)$.

Remark. (3) immediately implies the *Poincaré Lemma*: $\mathfrak{D}^p(D^n) = 0$ for $p > 0$, where D^n denotes the closed unit disc $\{\mathbf{x} \in \mathbf{R}^n | \|\mathbf{x}\| \leq 1\}$, since D^n is contractible to a point.

Remark. A drawback of the category of cornered manifolds is in that the boundary ∂M is not, in general, a cornered manifold. A drawback of the category of bounded manifolds is in that the direct product of two such manifolds is not always bounded (but cornered).

§ A2. The Morse Function

We refer the reader to Milnor [10] (up to p. 32) or Milnor [11] for the proof of the facts stated in this section. Hereafter, our category will be that of the compact bounded manifolds unless otherwise stated. A function $f\colon M \longrightarrow [a, b]$ is called a *Morse function on M* if f satisfies the following conditions:

(i) for a critical point $x \in M$ (i.e., a point at which $df = 0$), the Hessian Q_f at x (= the quadratic form given by the second derivatives $(\partial_{x^i}\partial_{x^j}f(x))$ of f at x) is nondegenerate,

(ii) $f^{-1}(a) \cup f^{-1}(b) = \partial M$,

(iii) ∂M does not contain a critical point, and

(iv) f has distinct values, $f(x) \neq f(y)$, at distinct critical points, $x \neq y$.

Any compact bounded manifold has a Morse function. Let $f\colon M \longrightarrow [a, b]$ be a Morse function. We identify the one-form df with a vector field $\operatorname{grad} f$ by means of a Riemannian metric. The integral curve $x(t)$ of $(-\operatorname{grad} f)$ through a given point is defined for $t \in [0, \infty)$ or else for $t \in [0, C)$ with $x(C) \in f^{-1}(a)$. The integral curves of $(-\operatorname{grad} f)$ define deformations of various submanifolds of M. If $f^{-1}([u, v])$, $[u, v] \subset [a, b]$, has no critical points in it then $f^{-1}([u, v])$ is diffeomorphic with $f^{-1}(u) \times [u, v]$. Suppose $f^{-1}([u, v])$ has a single critical point x in it. Let λ be the *index* of f at x (= the number of the negative eigenvalues of Q_f at x). Then $f^{-1}([u, v])$ is smoothly deformable to a compact cornered submanifold N which is obtained from the disjoint union $D^\lambda \times D^{n-\lambda} \cup f^{-1}([u, u + \epsilon])$ $n = \dim M$, for a small $\epsilon > 0$ by the attaching given by an imbedding $g\colon D^\lambda \times D^{n-\lambda} \longrightarrow f^{-1}([u + \epsilon, v])$ with $g(\partial D^\lambda \times D^{n-\lambda}) = g(D^\lambda \times D^{n-\lambda}) \cap f^{-1}([u, u + \epsilon])$. Thus $f^{-1}([u, u + \epsilon])$ has the same homotopy type as the topological space $D^\lambda \cup f^{-1}(u)$ attached at ∂D^λ. Hence λ is intrinsic in $(f^{-1}([u, v]), f^{-1}(u))$. For instance, if $\lambda = 0$ or n, $f^{-1}([u, v])$ is diffeomorphic with D^n. Denote by $\mu_f(M)$ the number of critical points f has. The Morse number $\mu(M)$ of M is by definition $\operatorname{Min}\{\mu_f(M) | f$ is a Morse function on $M\}$.

The above is summarized to:

Theorem A2. *For any compact bounded manifold M of dimension $n > 0$, there exists an $(n-1)$-dimensional compact submanifold V such that the following holds:*

(i) *if $\mu(M) = 0$, M is diffeomorphic with $V \times I$, where $I = [0, 1]$,*

(ii) *if $\mu(M) = 1$, M is smoothly deformable to a cornered submanifold $g(D^\lambda \times D^{n-\lambda}) \cap (V \times I)$ with $g((\partial D^\lambda) \cap D^{n-\lambda}) = g(D^\lambda \times D^{n-\lambda}) \cap (V \times I)$ where g is an imbedding and λ is an integer $\in [0, n]$.*

(iii) *if $\mu(M) > 1$, M is the union of two compact bounded submanifolds M_1, M_2 of dimension n, where $V = M_1 \cap M_2$, $\mu(M_1) = 1$ and $\mu(M_2) = \mu(M) - 1$.*

§ A3. The relative de Rham theorem

Let $H^p(M, N)$ be the pth singular cohomology group of (M, N) over the reals, defined from smooth singular chains (which is the same as the one defined from continuous singular chains). Then a homomorphism $\mathcal{S}(M, N)$: $\mathfrak{D}^p(M, N) \longrightarrow H^p(M, N)$ is well defined by taking the integrals of the closed p-forms $\omega \in \Omega(M, N)$ over p-cycles A in (M, N); in fact (i) if $\omega = d\theta$ with $\theta \in \Omega^{p-1}(M, N)$ then $\int_A \omega = \int_{\partial A} \theta = 0$ by $\partial A \subset N$ and (ii) if $A - \partial B \in N$ with $(p+1)$-chain B of M then $\int_A \omega = \int_{\partial B} \omega = 0$. (By digression we note that the definition of $\mathcal{S}(M, N)$ also applies to the category of the cornered manifolds; indeed, a singular simplex: $I^p \longrightarrow M$ belongs to this category but not to that of bounded manifolds.) $\mathcal{S}$ becomes a transformation from the functor $\mathfrak{D}^p$ into H^p in the obvious way; $f^* \circ \mathcal{S}(M', N') = \mathcal{S}(M, N) \circ f^*$ for f: $(M, N) \longrightarrow (M', N')$. Now we state our main theorem.

Theorem A3. *$\mathcal{S}(M, N)$: $\mathfrak{D}^p(M, N) \longrightarrow H^p(M, N)$ is an isomorphism for the category of the compact bounded manifolds.*

Proof. As is well known, the singular cohomology group satisfies the axioms of cohomology theory, (1)–(5). Hence we have the commutative diagram:

$$\begin{array}{ccccccccc}
\cdots \longrightarrow & \mathfrak{D}^{p-1}(N) & \longrightarrow & \mathfrak{D}^p(M, N) & \longrightarrow & \mathfrak{D}^p(M) & \longrightarrow & \mathfrak{D}^p(N) & \longrightarrow \cdots \\
& \downarrow \mathcal{S} & & \downarrow \mathcal{S} & & \downarrow \mathcal{S} & & \downarrow \mathcal{S} & \\
\cdots \longrightarrow & H^{p-1}(N) & \longrightarrow & H^p(M, N) & \longrightarrow & H^p(M) & \longrightarrow & H^p(N) & \longrightarrow \cdots,
\end{array} \tag{8}$$

which is exact in the rows upstairs and downstairs. By the five lemma, we will have $\mathcal{S}(M, N)$: $\mathfrak{D}^p(M, N) \cong H^p(M, N)$ for all p and all (M, N) provided we have

$$\mathcal{S}(M): \mathfrak{D}^p(M) \cong H^p(M) \quad \textit{for all } p \textit{ and for all } M. \tag{9}$$

We will prove (9) by double induction on $\dim M$ and the Morse number $\mu(M)$. When $\dim M = 0$, (9) is obvious from (5) for all p. Assume (9) is true for all p and all M with $\dim M < n$. Let M be a compact bounded manifold with $\dim M = n$. Suppose $\mu(M) = 0$. Then by Theorem A2, M is diffeomorphism with $V \times I$. Hence we have $\mathfrak{D}^p(M) \cong \mathfrak{D}^p(V \times I) \cong \mathfrak{D}^p(V) \cong H^p(V) \cong H^p(V \times I) \cong H^p(M)$ for all p by (3) and the induction assumption. Next suppose $\mu(M) = 1$. In view of Theorem A2, we need the commutative diagram:

$$
\begin{array}{ccccccccc}
\cdots \to & \mathfrak{D}^{p-1}(M_1 \cap M_2) & \to & \mathfrak{D}^p(M_1 \cup M_2) & \to & \mathfrak{D}^p(M_1) \oplus (\mathfrak{D}^p(M_2) & \to & \mathfrak{D}^p(M_1 \cap M_2) & \to \cdots \\
 & \downarrow \mathcal{S} & & \downarrow \mathcal{S} & & \downarrow \mathcal{S} & & \downarrow \mathcal{S} & \\
\cdots \to & H^{p-1}(M_1 \cap M_2) & \to & H^p(M_1 \cup M_2) & \to & H^p(M_1) \oplus H^p(M_2) & \to & H^p(M_1 \cap M_2) & \to \cdots
\end{array} \tag{10}
$$

which is exact upstairs and downstairs by the Mayor-Vietoris theorem for a cornered manifold $M_1 \cup M_2$, compact bounded manifolds M_1, M_2 and a smooth manifold $M_1 \cap M_2$. Substitute $g(D^\lambda \times D^{n-\lambda})$ and $V \times I$ both in (ii) of Theorem A2 for M_1 and M_2 both in (10) respectively. And (10) reads

$$
\begin{array}{ccccccccc}
\cdots \to & \mathfrak{D}^{p-1}(\partial D^\lambda) & \to & \mathfrak{D}^p(M) & \to & 0 \oplus \mathfrak{D}^p(V) & \to & \mathfrak{D}^p(\partial D^\lambda) & \to \cdots \\
 & \downarrow \mathcal{S} & & \downarrow \mathcal{S} & & \downarrow \mathcal{S} & & \downarrow \mathcal{S} & \\
\cdots \to & H^{p-1}(\partial D^\lambda) & \to & H^p(M) & \to & 0 \oplus H^p(V) & \to & H^p(\partial D^\lambda) & \to \cdots .
\end{array}
$$

In this diagram every manifold except for M has dimensions $\leq n - 1$. Thus five lemma gives (9) for all p and all M with $\dim M = n$ and $\mu(M) = 1$. Finally assume that (9) is true for all M with $\dim M = n$ and $\mu(M) < m$. Assuming $\mu(M) = m$, apply (iii) of Theorem A2 to M. And (10) implies (9) for this M. Thus Theorem A3 is proved.

Remark. We have proved the de Rham theorem for the category of compact bounded manifolds for the sake of simplicity. But it will be valid in larger or smaller categories. Here we only add that, if M is diffeomorphic with $M' - \partial M'$ for some compact bounded manifold M', the cohomology group $H^*_c(M)$ with compact support is isomorphic with $H^*(M', \partial M')$. And, if E is a metric vector bundle over a compact manifold, we have $H^*(E, E - E_{[0]}) = H^*(E, E - E_{(\epsilon)})$ for any $\epsilon > 0$.

BIBLIOGRAPHY

[1] M. F. Atiyah and I. M. Singer, *The index theorem of elliptic operators*. III, Ann. of Math. (2) 87 (1968), 532–546.

[2] R. Bott, *Vector fields and characteristic numbers*, Michigan Math. J. 14 (1967), 231–244. MR 35 #2297.

[3] R. Bott and S. S. Chern, *Hermitian vector bundles and the equidistribution of the zeros of their holomorphic sections*, Acta Math. 114 (1965), 71–112. MR 32 #3070.

[4] S. S. Chern, *On curvature and characteristic classes of a Riemann manifold*, Abh. Math. Sem. Univ. Hamburg 20(1965), 117–126. MR 17, 783.

[5] D. Husemoller, *Fibre bundles*, McGraw-Hill, New York, 1966. MR 37 #4821.

[6] S. Kobayashi, *Theory of connections*, Ann. Math. Pure Appl. (4) 43 (1957), 119–194. MR 20 #2760.

[7] ———, *Fixed points of isometries*, Nagoya Math. J. 13 (1958), 63–68. MR 21 #2276.

[8] S. Kobayashi and K. Nomizu, *Foundations of differential geometry*, Vols. 1, 2, Interscience Tracts in Pure and Appl. Math., no. 15, Interscience, New York, 1969.

[9] J. Leray, *Le calcul différential et intégral sur une variété analytique complexe*. (*Problème de Cauchy*. III), Bull. Soc. Math. France, 87 (1959), 81–180. MR 23 #A3281.

[10] J. Milnor, *Lectures on the h-cobordism theorem*, Princeton Univ. Press, Princeton, N. J., 1965. MR 32 #8352.

[11] ———, *Morse theory*, Ann. of Math. Studies, no. 51, Princeton Univ. Press, Princeton, N. J., 1963. MR 29 #634.

[12] T. Nagano, *Transformation groups with $(n-1)$-dimensional orbits on noncompact manifolds*, Nagoya Math. J. 14 (1959), 25–38. MR 21 #3513.

[13] ———, *Homotopy invariants in differential geometry*. I, Trans. Amer. Math. Soc. 144 (1969)

[14] M. S. Narashimhan and S. Ramanan, *Existence of universal connections*, Amer. J. Math. 83 (1961) 563–572. MR 24 #A3597.

[15] H. Samelson, *On de Rham's theorem*, Topology 6 (1967), 427–432. MR 35 #6156.

[16] ———, *On Poincaré duality*, J. Analyse Math. 14 (1965), 323–336. MR 31 #5203.

[17] E. H. Spanier, *Algebraic topology*, McGraw-Hill, New York, 1966. MR 35 #1007.

[18] R. Thom, *Espaces fibrés en sphères et carrés de Steenrod*, Ann. Sci. Ecole Norm. Sup. (3) 69 (1952), 109–182. MR 14, 1004.